REMEMBERING
Randall

A Memoir

OF POET, CRITIC, AND TEACHER
RANDALL JARRELL

BOOKS BY RANDALL JARRELL

POETRY

The Rage for the Lost Penny (in Five
  Young American Poets)   1940
Blood for a Stranger   1942
Little Friend, Little Friend   1945
Losses   1948
The Seven-League Crutches   1951

Selected Poems   1955
The Woman at the Washington Zoo
  1960
The Lost World   1965
The Complete Poems   1969

ESSAYS

Poetry and the Age   1953
A Sad Heart at the Supermarket
  1962

The Third Book of Criticism   1969
Auden, Kipling & Co.   1980

FICTION

Pictures from an Institution   1954

CHILDREN'S BOOKS

The Gingerbread Rabbit   1964
The Bat-Poet   1964

The Animal Family   1965
Fly by Night   1976

TRANSLATIONS

The Golden Bird and Other Fairy
  Tales of the Brothers Grimm
  1962
The Rabbit Catcher and Other Fairy
  Tales of Ludwig Bechstein   1962

"The Three Sisters," Anton Chekhov
  1969
Snow White and the Seven Dwarfs of
  the Brothers Grimm   1972
Faust: Part I   1976

ANTHOLOGIES

The Anchor Book of Stories   1958
The Best Short Stories of Rudyard
  Kipling   1961
The English in England (Kipling
  stories)   1963

In the Vernacular: The English in In-
  dia (Kipling stories)   1963
Six Russian Short Novels   1963
Randall Jarrell's Letters Edited by
  Mary Jarrell   1985

With Kitten

REMEMBERING

Randall

A Memoir

OF POET, CRITIC, AND TEACHER
RANDALL JARRELL

MARY VON SCHRADER JARRELL

Perennial

An Imprint of HarperCollinsPublishers

Parts of several essays were previously published in *Randall Jarrell: 1914–1965:* Farrar, Straus & Giroux, also *Parnassus, Harper's Shenandoah, USA* and *The Columbia Forum.* All Randall Jarrell poems and excerpts are from *The Complete Poems:* Farrar, Straus & Giroux.

Quote from *Desolation Angels* by Jack Kerouac on pages 42–43 reprinted by permission of Sterling Lord Literistic, Inc. Copyrighted by Jack Kerouac, 1965. Renewed 1993.

A hardcover edition of this book was published in 1999 by HarperCollins Publishers.

HarperCollins books may be purchased for educational, business, or sales promotional use. For information please write: Special Markets Department, HarperCollins Publishers Inc., 10 East 53rd Street, New York, NY 10022.

First Perennial edition published 2000.

Designed by Stanley S. Drate/Folio Graphics Co., Inc.

The Library of Congress has catalogued the hardcover edition as follows:

Jarrell, Mary.
    Remembering Randall : a memoir of poet, critic, and teacher
Randall Jarrell / Mary von Schrader Jarrell.—1st ed.
        p. cm.
    ISBN 0-06-118011-4
    1. Jarrell, Randall, 1914–1956—Marriage. 2. Poets, American—20th century Biography. 3. English teachers—United States Biography. 4. Critics—United States Biography. 5. Jarrell, Mary. I. Title.
PS3519.A86Z7   1999
811'.52—dc21   99-13048
[B]

ISBN 0-06-118013-0 (pbk.)

00 01 02 03 04 ❖/RRD 10 9 8 7 6 5 4 3 2 1

To

SUSAN and FRED CHAPPELL

&

BETTY and ROBERT WATSON

There is no safer way to avoid the world
than through Art;
there is no safer way to be linked
to the world than through Art.

Johann Wolfgang von Goethe

Contents

GRATEFUL APPRECIATION

For the benevolent counsel of my daughters, Alleyne Garton Boyette and Beatrice Garton Garbett.

For the companionship of my spiritual brothers and sisters: Reverend John Akers, Jean-Paul de Caussade, Jane Darnell, Madeleine L'Engle, Reverend Terry Fullam, Jan Hensley, and Paul of Tarsus.

And finally for Ted Russell's incomparable photographs, and for my accomplished and ever-accommodating amanuensis, Betty Bullington.

M. von S. J.

INTRODUCTION

In a letter to his red-haired Viennese friend, Elisabeth Eisler, Randall prided himself on being "quite optimistic, mostly in order to save bother: I accept, dismiss, and forget about bad things that happen as quickly and well as I can." I guess one of the great principles of my life is: *O, don't bother, forget about it . . .*" Saving bother coupled with my own great commitment to William James's "Wisdom is knowing what to overlook" spared us many a squabble. Still, on rare occasions a matter came up between us that was not . . . was *not* overlookable. Such was the case in the domestic havoc Hannah Arendt unknowingly caused us over my fury at Randall's intention to dedicate *Pictures from an Institution* (the darling-of-our-hearts) to Hannah instead of to me.

At that time, I only knew Hannah, whom he came to know over their lunches at Sarah Lawrence, through Randall's warm praise for this German-born political theorist and author of *The Origins of Totalitarianism*. In defending the dedication, Randall explained repeatedly Hannah's contributions to his mocking comparisons of American education compared to European, which of course was the underlying content of *Pictures*. He spoke fondly of Hannah's belatedly according him the privilege of first-name status that Germans are so cautious to grant. Finally, he reminded me that Hannah was the model for his character Irina in the book.

After a half-eaten lunch when I still objected, Randall admitted he had promised her the dedication long before we met and that he could not go back on it now.

"Ask me anything but that," he pleaded. "And any books to come. But be reasonable, beloved, can't you?"

"I know, I know," I countered. "But I would never dedicate a book of *mine* to another man no matter what. It's not our 'group of two.' It's you and Hannah. And I feel slighted. And it hurts."

I left him talking to the air and fell facedown on our bed crying into my pillow. When Randall came in he sat on the side of the bed and gently smoothed my shoulders and crooned helplessly, "There, there, baby. *Please* don't cry. *Please.*" In a strained voice he went on, "We *are* our group of two. Nothing can change it. *EVER.* Cross my heart." Picturing that, I turned my red and red-hot face to him and felt a foolish smile trying to come.

Not wasted on the former psychology major who quickly dabbed my eyes with the bedspread. Heaving a sigh of relief he slapped his brow and mumbled, "Oh, I'm so dumb." And sparks of optimism returned to his eyes. "What about this, baby? If we dedicate the silly book to Mary *and* Hannah? How would that be?"

I nodded and sat up in bed entirely appeased, and Randall carried on chattily as if nothing had happened, "I have a surprise, you dear, funny thing. Guess what, there's a Garbo rerun playing up in Balboa. Why don't you put on that dovey bare-shouldered paisley and we'll go out in the great world and laugh and play, Eh? Dinner, too."

M. von S. J.

I

Sometimes a poem comes to me—I do what I can to it when it comes—and sometimes for years not one comes. During these times the only person who helps much is my wife: she always acts as if I'd written the last poem yesterday and were about to write the next one tomorrow. While I'm writing poems or translating Faust *I read what I have out loud, and my wife listens to me. Homer used to be led around by a little boy, who would listen to him; all I can say is, if Homer had ever had my wife listen to his poems, he would never again have been satisfied with that little boy.*

RANDALL JARRELL

From the National Book Award acceptance for
The Woman at the Washington Zoo

Celestial Navigation Tower Operator

U. S. ARMY AIR FORCE

Ideas and Poems

I REMEMBER A SPARKLING low-humidity Sunday in Nashville at the country club with Randall's rich Uncle Howell Campbell. Uncle Howl was a candy manufacturer of Bell-Camp Chocolates and Goo-Goo Bars. After Randall's parents were divorced and he and his mother and brother left California, Uncle Howl took over most of their support, even to paying Randall's day student tuition at Vanderbilt. So, while we dined on congealed salad, batter-dipped, deep-fried okra, and pecan pie still warm, Uncle Howl was quizzing Randall on how he got ideas for poems. Randall answered him haltingly, "Gee, I don't know . . . Some of it is luck. You have so little control over what you write . . . Ideas come to me . . . Or they don't. It's hard sometimes . . ."

And Uncle Howl broke in, "Now wait a minute, son. Hold it. I don't go with that. Why I don' have any trouble gittin' ideez. Plenny ideez aroun'. Lemme tell ya. Nex' time you give outta ideez, Ran, gimme a ring. Colleck . . . Why your Uncle Howl c'n give you alla ideez you'll ever . . ." Randall turned to me with a look that said, "Uncle Howl thinks one can crank out poems like Peppermint Patties."

Actually, they were fond of each other, and Randall's uncle just wanted him to succeed. In a well-intentioned way, Uncle Howl sometimes suggested ideas for poems "that will sell, son." And although Randall couldn't-didn't use them, he did use Uncle Howl's speech rhythms and businessman attitudes in a dramatic monologue called "Money."

The hero of "Money" was not Uncle Howl. He was an imaginary millionaire on milk toast who had outlived two wives and outbid Clay Frick for a "dirty Rembrandt bought with dirty money." Randall invented him from ideas he took from tycoons in the twenties. The fact that they talked Uncle Howl's language was a happy convenience for the poet to use the way a painter will use his mistress to pose for the Madonna. The kernel of "Money"—what von Hoffmansthal calls the "pure, underlying *poetic content*"—came from Randall's anticapitalist and Marxist leanings in the thirties.

Back in Nashville, with his book open at the "Money" page, Uncle Howl would whap it and declaim in heavy iambs, "Ran got it all from *me*! I give'm those ideez! No mam! I won't quit braggin'! Got me somep'n to brag on! Shoot! Ran don't *know* any folks 'cep' me's *got* money!"

Of course, no one gives ideas to a poet. What happens is that we all give off ideas, and if the poet perceives one that solves some part of his poem, he takes it. Very different from being *given* it. Only the writer knows—never his Uncle Howl—if he is drawn to the idea; if it has potentiality; and if it should be left as is, sawed in half, or melted down, the better to metamorphose it into the substance of poetry.

When, with luck, the good ideas came—even the very good ones that tennis-player Randall said were "right up my alley"—it was no guarantee a poem would come; and, if it did, if it would stay. Many a poem off to a robust start took leukemia, weakened, and died, so that we were wary of talking about partial poems. Better to wait and see if those ideas would fit and link and multiply. If the poem would "click" the way Frost said it must, "as if you'd turned a key in a lock—and you can only turn it once." How long that would take, we never knew.

What is mysterious about the artist's ideas, according to Rilke, is that while one is the servant of one's art one "cannot evoke it. I am not an 'author' who '*makes*' books. The *Elegies* (if it is granted me to complete them) will depend on an internal organization." A state of "readiness to transform receptivity into productivity." In trying to explain to his fond Merline why they cannot be together in the near future Rilke continues, "It is for this 'readiness' that I struggle now and no one must touch me or shake me, for like the formation of a crystal it depends on the most distant influences, which can only reach us when we are standing inside the constellation, undisturbed." "Inside a constellation undisturbed," Randall murmured. "What bliss. What prose! Who else but Rilke?" Whom Hannah taught him to pronounce "Ril-ka."

With "belief and patience," Rilke waited out a ten-year lapse after starting the *Elegies*. In a castle here and a Schloss there he occupied himself with such prose in his voluminous correspondence: and then "the miracle" of his readiness took place. It was given him to complete ten *Elegies* and fifty-five *Sonnets to Orpheus* in a hurricane of writing day and night and "with food hardly to be thought of."

Kipling had no such barren lapses but heaped credit for his steady flow of ideas on his inspirational Daemon, a Daemon in the Greek sense that he looked to and trusted beyond himself. Assisted by his Daemon, Kipling created an oeuvre of twenty some volumes of prose and poetry, which won him a Nobel Prize for Literature and—presumably—his final resting place in the Poet's Corner in Westminster Abbey.

So much for miracles and Daemons. The fact is we never wore out Uncle Howl's question, "Where do a poet's ideas come from?" Freudian Randall allowed it was the Unconscious, and my mother sent us her down-to-earth contribution to the subject in the form of a news clipping in a bread-and-butter letter after a visit to us. I had a hunch it wouldn't suit Randall and slipped it back into the envelope, but took it out and read this aloud when he asked me to:

> The creative mind in any field is not more than this: a human creature born abnormally, inhumanly sensitive. To him a touch is a blow, a sound, a noise, a misfortune is a tragedy, a joy is an ecstasy, a friend is a lover, a lover is a god . . . Add to this cruelly delicate organism the overpowering necessity to create, create so that without the creativity of music or poetry or books or buildings or something of meaning his very breath is cut off from him. He must create, must pour out creation. By some strange, unknown, inward urgency he is not really alive unless he is creative.

"Huh! Tell *me*!" Randall said derogatorily. "And who, pray, wrote that cant? Not Freud."

"Not Proust."

"Silly thing! Who?"

"Pearl Buck."

"*Good Earth*, Pearl Buck? R-dick-a-lus."

"I take it you haven't read her?"

"Nope. Not even skimmed."

"And why not? Pray."

"Well, von S., if you must know, and don't I tell you everything?" he said half in truth, half in jest. "Any more of *that* would corrupt my prose style. I'm going to kiss you."

I am remembering now when Randall and I first laid eyes on each other at the Rocky Mountain Writers Conference at the

University of Colorado in Boulder. It was summer vacation for him from his all-girl classes at the Woman's College of North Carolina in Greensboro. It was also summer vacation for me, from Emerald Bay, Laguna, California, in that my two daughters by my former husband were in Europe biking with their architect father on a castle and cathedral tour.

On a hot, class-free Saturday in July we spent the day in Denver trying to buy Proust's two-volume *Remembrance of Things Past* for me and the latest Bruno Walter recording of Mahler for Randall. I, the unwalker of the world, was continually outpaced by athletic Randall in his brown and white saddle shoes, but was cheered on by his singsong "Just one more block? Pretty please? Who knows what we might find, eh?"

We found both and *finally* paused for refreshment at Brown's Hotel. Randall downed one tumbler of orange juice after another and I cooled off with a long-lasting mint julep. Randall did the talking and I did the listening and he divulged his angst over the poet's lack of idea-control combined with the nightmare of losing one's poems entirely. "It's something that happens to poets," he said. "They stop writing poems. It happened to Eliot. It happened to Tate and Mr. Ransom [John Crowe]. It's happening to Dylan Thomas."

The next day we were proofreading the long galley sheets for his *Seven-League Crutches*. "It could be my last book." Randall sighed. "You never know . . . I haven't had an idea for a poem in months." A day or so after that, on a walk near a dry stream bed, I picked up an odd, vermiculated, worn-down rock and cried out, "A meteorite. Whatever that is . . ." "A meteorite?" Randall said, playing along and ever celestially oriented. "A meteorite? Oh, it's a porous, burned maybe, rock . . . like this one . . . that breaks off in space . . . and space is full of them . . . and may catch fire from its proximity to stars on its way down

to the planet to our very feet. Why don't you take it back to Laguna, best beloved?"

"*Beloved*!" I said, "As in Kipling?"

"So *clever*!" Pats of approval. "Words fail me."

A week later Randall had written a poem called "The Meteorite." It was sentimental—what critics call "slight" if they
mention it at all—but it was a *poem*.

In it Randall turned me into a star and turned himself into
a meteorite "half iron and half dirt" to whom the star bends
and puts to her lips and breathes "upon it until at last it
burned." The poem was only six lines but the last line,
"Breathe on me still, star, sister," held the pure *underlying* poetic
content, in the von Hoffmansthal sense, of *constancy*. Dear to
Randall as was Rilke's theme of change, dearer still was his
own dream and hope for un-Change: for Permanence, in that
Permanence is what we all want when we can love and can be
loved; Change is what we want when we cannot.

For me to be, as that star, looked up to and almost worshipped, was to be loved, as Yeats said, "in the old high way." In
return for this, to "bend to *one*" steadily—faithful as a star—
was . . . was . . . But I will say it, was all I wanted.

The Meteorite

> *Star, that looked so long among the stones*
> *And picked from them, half iron and half dirt,*
> *One; and bent and put it to her lips*
> *And breathed upon it till at last it burned*
> *Uncertainly, among the stars its sisters—*
> *Breathe on me still, star, sister.*

Naturally, anyone close to the poet becomes a source of
ideas; it is an occupational hazard a mate must accept. When I

see my virtues exalted in Randall's poems I think, "So! All in all I wasn't so bad for him." And when my vices were made public, I told myself, "It's only symbolic, Mary—everybody's like that."

Having one's chat and one's dreams and tears used: being used oneself by the artist—not because you are You, but because you are useful and usable is no mean existence for a wife. I soon rid myself of the notion that I *knew*—like Uncle Howl—what would *do* for a poem. My aim was to make available—but not suggest—all such sweet and sour samples of the human condition that came my way. What I did was put *my* sensibility at *his* disposal. It was a joyful assignment for I seem to be a born confidante with a parrot's ear for dialogue and a gossip's nose for pertinent anecdotes. When I found Randall, I found what I was programmed for. And—apart from our bewildering sad phase toward the end—I reveled in my years with him: fetching ideas, always fetching and always in the dark about what he would keep and what he would turn it into.

Once inside the constellation and at the ready to produce, Randall was a rapt servant of his work. Oblivious of "sources of ideas," he would happily quote Blake's "I will not reason and compare. I will create." An example of just how oblivious he could be occurred in *The End of the Rainbow*.

After a long afternoon's chat about my grandmotherly New England Cousin Bertha and Su-Su, her Pekinese, Randall wrote a novelette-of-a-poem about an elderly New England maiden lady he named Content and endowed with a Freudian-tinged case history. He planted Content with her porringers and ancestral samplers in Laguna, which gave him the opportunity to poke fun at some Southern California customs.

All in all I was enchanted by what a creative mind can create from a few scraps of conversation. I was enchanted with her landscapes and seascapes, her turquoise and plate-glass store for art supplies, turquoise and plate glass being inspired by my own hillside home. In an arm-in-arm walk along the seashore, I was lavishing praise on *The End of the Rainbow*.

"You even kept Su-Su's real name in it. I love it. It's exactly right, isn't it?"

Randall looked almost dumbfounded. "Su-Su was actually the name of *Cousin Bertha's* dog? I thought I invented Su-Su." I hugged his arm and said, "Of course—just like Vermeer invented the checkerboard floor."

After Boulder, when I got home to Laguna, Randall sent me weighty literary bundles of Kafka, Rilke, Eliot, Lowell, Bishop, and Berryman; Auden, Chekhov, Ransom, Frost; Isak Dinesen, Colette, Christina Stead, B. H. Haggin, Sir Kenneth Clarke, Peter Taylor and, of course, Jarrell. After those came Stendahl's *Charter House of Parma*, all the recorded Mahler of the time and two albums of Learn German records "to play when you don't have anything to do."

We wrote each other every day while Randall taught at Princeton and we waited for his divorce. Due to his choice, Randall's letters to me were addressed to Miss Mary von Schrader; and once or twice to Emerald Bog instead of Emerald Bay—due to my unruly penmanship.

We used air mail and special delivery but sometimes my letters to him were delayed and Randall wrote me, alluding to Shakespeare's Bottom the Weaver, "Here Bottom and his band are running the mail system. And I've been to the Princeton mail room (where a mild old gentleman with gray hair and a brown toupee told me at ten minutes length everything about mail so as to prove I *couldn't* have any), and the main post office where the 'mail carriers don't sort mail until after they come back at 3:30 and you would have to have the postmaster's consent to intercept mail.' What Princeton needs is an Institute of Advanced Postmen." Another letter closed, "I'd better get this off and not give Bottom a chance." Eventually, "old Bottom" became the family scapegoat. When the toast caught fire or the bank put us on Ready Reserve or the car ran out of gas, not to mind. Not to swear or point fingers. One of us would usually—almost always—say, "Just old Bottom."

On a happier note, when the California mail was on schedule Randall began decorating the outside envelope with two stars inside the curve of a crescent moon and closed one letter saying, "How do you like my drawing of the crescent moon and two stars? I think it is a pretty emblem for us. I like it being physically impossible though spiritually *so*."

"The Three Bills," a much later and more sophisticated poem, came to Randall in a morning. It was about real people in a real incident that needed no disguises; simply needed him to write it down. Like Uncle Howl's "Money," it was about money but a far cry from Marxism.

As I recall, we were in New York back when Altman's was on Fifth for an interview with Randall on Channel 13. We were staying at the Plaza, which we had acquired a taste for staying in and where, in those days, they gave discounts to academics.

We liked their leisurely half-empty elevators where one ascended to one's floor with a couple speaking French. We liked our breakfasts in the Edwardian Room, waited upon by trained professionals, where one could loiter over one's sweet rolls adorned with pecans a generous inch long and could indulge in the Plaza's pastime of Seeing and Being Seen.

The morning of "The Three Bills" we were discreetly eyeing an elegant family whose small daughters wore velvet smocks with Velázquez lace collars and watching the demure and beautiful and young wife lighting her husband's cigar.

"Wealthy Brazilians," I whispered to Randall.

Then, over a steaming cup of fresh coffee, we planned our morning: Randall to Goody's for records and I to Bergdorf's where once a year I could find on sale one item I could afford. We agreed to meet later in our room. Suddenly, by a trick of acoustics, I was tuned in to a bizarre conversation behind us. In a moment Randall said, "Old pet, are you listening to me, or not?" And I said, "Not, dear one. I'm eavesdropping on something beyond belief. I'll tell you about it when they finish."

"Where *are* those people?" Randall said softly. And we turned around just in time to locate them as they were leaving.

Bill One—who'd been complaining about a plantation someone had loaned her in the Caribbean "at the end of nowhere"—was a compact, old-ish matron in a dowdy crepe dress and wearing a sizable diamond bar pin. "Third generation Social Register," I whispered as she went past. And Randall shushed me, fearing she'd heard.

Bill Two, whom the maitre d' addressed as Sir Ian, was London tailored, though it did not hide the fact there must have been under his suit a doughy white body just like his doughy white face.

Bill Three was his wife, a lady still fine and fair by Elizabeth Arden standards, but whose face—by ours—had a dire future. She looked at us sadly as she passed, and we looked sadly back, mindful of her words when Sir Ian was absent from the table. "We can't stay anywhere. We haven't stayed a month in one place for the last three years. He flirts with the yardboys and we have to leave."

Later we set out in good spirits and kissed goodbye at the Plaza fountain. Across the street at Bergdorf's I found a lightweight pearly-beige dress: one-piece, flared, nearly sleeveless and unmistakably mindful of the kind of day dresses Jackie Kennedy wore with her pillbox hats.

Ultimately that dress was immortalized in Randall's poem "A Man Meets a Woman on the Street," where he speaks of it as his wife's "new dress from Bergdorf's . . ."

And:

> *Women were paid to knit from sweet champagne*
> *Her second skin: it winds and unwinds, winds*
> *Up her long legs, delectable haunches,*
> *As she sways, in sunlight, up the gazing aisle.*

When I returned to the room I found Randall propped up in the cushions of an unmade bed littered with Plaza stationery. He greeted me with, "Best beloved, old Ramble has a surprise for you. Oh, yes. Oh, yes. Sit ye down and get ye comfy." Instead of showing me a haul of Strauss and Schwarzkopf he started to read me "The Three Bills," the only poem he ever finished at one sitting.

Then, as often happened when I heard a poem for the first time, the Mary Jarrell that I was—faculty wife, mother, homemaker, American citizen in New York—was transcended as Randall's voice lifted another Me to a wholly new time and place, where strange, yet dimly familiar persons added their joys to my joys and exchanged their woes with me.

Three Bills

> *Once at the Plaza, looking out into the park*
> *Past the Colombian ambassador, his wife,*
> *And their two children—past a carriage driver's*

Rusty top hat and brown bearskin rug—
I heard three hundred-thousand-dollar bills
Talking at breakfast. One was male and two were female.
The gray female complained
Of the plantation lent her at St. Vincent
"There at the end of nowhere." The brown stocky male's
Chin beard wagged as he said: "I don't see,
Really, how you can say that of St. Vincent."
"But it is at the end of nowhere!" "St. Vincent?"
"Yes, St. Vincent." "Don't you mean St. Martin?"
"Of course, St. Martin. That's what I meant to say, St. Martin!"
The blond female smiled with the remnants of a child's
Smile and said: "What a pity that it's not St. Kitts!"
The bearded male went for a moment to the lavatory
And his wife said in the same voice to her friend:
"We can't stay anywhere. We haven't stayed a month
In one place for the last three years.
He flirts with the yardboys and we have to leave."
Her friend showed that she was sorry; I was sorry
To see that the face of Woodrow Wilson on the blond
Bill—the suffused face about to cry
Or not to cry—was a face that under different
Circumstances would have been beautiful, a woman's.

The Complete Poems

My first awareness of the present again was the sight of Randall's pure and exalted look at the ending just before the poem's spell broke. And, once again, it crossed my mind before either of us spoke how *different* the world is to the poet, and to the nonpoet. Still fresh, the incident in the Edwardian Room had seemed no more than a little anecdote to take South, tell a friend or two, and forget. Randall had seen a latter-day "Money" in it, as told by the tycoon's trust fund heirs whom dollars had turned into dollars, but whom he had turned into poetry.

Without Randall my ideas, my relatives, my eavesdropping and all else that I fetched up would have remained raw ore. In passing through his unique atmosphere they were incandesced and metamorphosed from something half iron and half dirt to art.

II

I rarely feel happier than when I am in a library—very rarely feel more soothed and calm and secure. And there in the soft gloom of the stacks, a book among books, almost, I feel very much in my element—a fish come back to the sea; a baby come back to the womb. I like libraries so much I feel depressed that my cat can't check out books, too.

RANDALL JARRELL

From a talk to the American Library Association

Jarrell, Mary, Alleyne (11) and Beatrice (8): Laguna

BOB WILLOUGHBY

Libraries

"WHEN I WAS A BOY," Randall continued with his librarian audience, "I never had more than a sweet roll and a half pint of milk for lunch because I spent half my lunch money on street-car fare to the Carnegie Library in uptown Nashville."

He got his first library card at age six, he told me, and for years, Randall waited there after school for his mother to pick him up after work. Along about age twelve, Randall said, he developed a crush on a young, red-haired lady assistant in circulation, which—I found later—gave him a weakness for young, red-haired ladies for the rest of his life.

Our first date was an evening in a university library, and during our lifetime together, his last errand on a moving day was to a library to return his books and his first errand in the new town was to a library for a card. Never mind establishing the water and lights.

I brought two dearly loved daughters into our household in Greensboro: Alleyne, eleven, and Beatrice, eight. Randall brought dozens of hand-carried seventy-eight records, six engraved tennis trophies, and Kitten, who was not a kitten but a huge, dearly loved, and immensely intelligent black cat.

Spring of 1956 was idyllic, idyllic meaning the kind of dog-woods-and-daffodils month all over town that department heads hoped for when they scheduled April interviews for prospective faculty. No nasty cold snaps threatening to nip all the buds in the bud. That long series of mild days inspired the Modern Poetry class to meet under the Japanese cherry trees while Randall read Eliot's "Prufrock" aloud, accompanied by the courting arias of a resident mockingbird.

It was also Kitten-weather when Kitten got inspired to spring up the pecan tree near our driveway and stretch out along a leafy branch hiding from Randall and pretending he couldn't get down. Randall, pretending he didn't know where Kitten was, wandered around the yard softly calling, "Kitten? Kitten? Where is my magic Kitten?" After a while, as if by chance, he would look up in the pecan tree and I'd hear his joyful cry, "Oh, *there* you are! *Clever cat.*" And he'd reach up, as both of them knew he would, and pluck Kitten off the branch to ride on his shoulder and purr in Randall's ear as they entered the house.

According to one of Alleyne's friends, "Nineteen hundred fifty-six was the year Venus was so close to the earth." True? Or? I do remember Venus as being very bright and usually the first star of the evening. Weatherwise that spring was just right for our top-down drives in the Mercedes to see the newest Ingmar Bergman film in Chapel Hill, and just right for all the tennis Randall and Parole Officer Carrigan had time for.

An additional benevolence was that Randall had so enamored his Modern Poetry class of "Prufrock" that they voted to forfeit Auden and Yeats in order to devote the rest of the semester to Eliot. They even renamed the Modern Poetry class "the Eliot Girls."

Behind his fervor was Randall's conviction that his critical essays *Poetry and the Age* was "a flawed masterpiece" in that it had lacked Eliot. This had grieved him to the point that all he wanted to write now was to remedy that, but his novel *Pictures from an Institution* had intervened. Now, in the spring of 1956, as he focused daily on Eliot to prepare the Eliot Girls' lectures, old and new ideas were crowding his mind, and the summer vacation beckoned him to honor Old Possum with the kind of in-depth essay he had written for Whitman, Graves, and Frost. I could see an air of buoyancy about him.

Then out of the blue came Librarian Quincy Mumford's invitation—promoted by poets Allen Tate, Red (Robert Penn) Warren, and Cal Lowell—for Randall to serve the next two years as the Poetry Consultant (now Poet Laureate) at the library of libraries—the Library of Congress. "Not only were we happy," as Randall said Kipling said of his childhood in India, "we knew we were happy."

After the required personal interviews in Washington, Randall wrote Roy Basler, Head of Reference and majordomo of the Library's cultural activities:

"I trust the forms were complete and the fingerprints clear—the police lieutenant seemed very professional. The big NON-SENSITIVE on the forms made me feel good, as if I'd at last made the football team in the line at that. I'm glad to know that the general reaction was one of approval. As you say, 'you never know.'"

Basler's "you never know" was an allusion to the public hubbub over the Library's forced rejection of poets Ezra Pound's and William Carlos Williams's appointments to the Chair (so-called at times) due to McCarthy-ish questions of "security." Karl Shapiro, Pulitzer poet and WWII veteran, passed muster in the interim period and regaled other possible consultantship candidates—like Randall—saying former Librarian Luther Evans's first words of greeting to him were, "Shapiro, we don't want any Commies or cocksuckers in this Library."

By the time the milder-spoken Mumford replaced Evans and the Senate had censored and condemned McCarthy's witch-hunting in Hollywood, academe, and all departments of government—even the military—McCarthy-itis was dying out, but as Roy implied, McCarthy adherents were still alive and lively.

Randall's appointment made news in the Washington, New York, and Greensboro papers, and friends and relatives overwhelmed us with clippings. I booked a mid-August moving date with Bekins and started collecting empty liquor cartons from the ABC store for packing our long-stemmed green Rhine wineglasses and other fragiles to personally transport in the car. We had subscribed to the *Washington Post* for Randall to read up on his Redskins and for me to find my dream house in Georgetown for rent.

We were in a blissful Washington/Eliot bubble until the end of the semester when our friend Roy made a frantic phone call saying Randall's appointment was in dire jeopardy for—yes—*security* reasons.

According to Roy word had come to him from "a prominent North Carolina family" that Randall was a "card-carrying Communist" who wrote for such "pinko publications" as *Partisan Review*, *The New Republic*, and *The Nation*. This source also claimed Randall had Marxist friendships from his years at the University of Texas. And finally, they protested that such a "prestigious government position" should be given to him and paid for with taxpayers' money. Though Roy decently withheld the author's name on this threatening letter, Randall and I knew whose fingerprints were all over it. Who else would it be other than a certain student in the Eliot class who was outraged by receiving a C on her report card instead of a B? Who with her family fought for Randall to change her C? On being firmly advised that her work was not comparable to the B students' work and it would not be honorable for him, or fair to them, to accommodate their wishes, she and her family had departed in high dudgeon. Every professor's nightmare.

Shortly after this Randall was treated to a spate of unsigned postcards rebuking him for his "Castro-beard," his "un-American imported car" and his "German wife." He deftly pitched them into the trash, deeming them unworthy to speak of even between ourselves.

Roy was distraught over the embarrassment this would cause the Library, and Randall was solemnly assuring him he'd only joined two organizations in his life at this period, Phi Beta Kappa and the Air Force, and the only two cards he'd ever carried were a driver's license and a library card. That didn't do it.

What was required, Randall told me after the phone call, was a character recommendation from Chancellor Graham at the university, vowing that Randall was Communist-free.

So Randall was embarrassing to the Library? What about Randall's embarrassment in needing this crucial clearance from the university chancellor, whom Randall was leading a faction to remove? I began dreading the letters and explanations to the clipping senders and to our faculty friends. Even worse was the panic for our future. What if Chancellor Graham took this as an opportunity to spoil Randall's "prestigious appointment" and, possibly, terminate his cherished teaching position? Could this student's wretched C rob us of our two years in Washington? Our livelihood from Randall's teaching? Our friends, who might regard us as untouchables? I knew better than to add my anxiety to his and thought to calm my nerves by cooking our favorites: ratatouille; grainy, sugarless cornbread oven-baked in a preheated iron skillet; and Hannah's Röte Fruchte, port wine gelatin crammed with red raspberries, red strawberries—frozen if need be—and red pitted Bing cherries with real whipped cream on top. Alas, Randall could ingest nothing but tea and crackers and could do nothing but play tennis—doubles, even—to wait this out. The suspense was, in Randall's word, "hellish." In the end our friend Dean Mossman, the Chancellor's assistant, worded the letter and got his signature. The Library appointment was saved, and not a hint of its threat ever made the news.

The Eliot Girls gave us a farewell party, and Randall began carting books back to the university library, and the Bekins people sent us a driver who looked to Randall like Red Sox hitter Ted Williams. The only affordable Georgetown house I found was too small, on a heavily trafficked street too perilous for a cat, and while I stood in the living room talking to the agent, my French heels were perforating the aged wooden floor. We settled for a twenties Dutch colonial near the Maryland border.

We followed Ted Williams in the Mercedes crammed with Randall's best racket in its frame, the von S. stemmed Rhine wineglasses in a Jack Daniels carton under my knees, and Elfie panting in my lap. Kitten was deceased. In Alexandria the Bekins van pulled to the roadside, as the driver wanted Randall to take the lead for the rest of the way. Stooping to talk through the Mercedes window, he said, "Alla drivers hate these D.C. moves. Now, take New York. Piece of cake. Laid out on a grid. But D.C. is a bitch. Wherever you go you get lost and wind up back at the Capitol."

Weary, Randall said, "Hmm, I never thought of that. Rather like a spider web?"

"More like a octopus," the driver said, flashing a tired Ted Williams smile, and Randall led us to the house in Chevy Chase.

It was in a Calvin Coolidge neighborhood on a tree-arched street named Jenifer. Veteran Washingtonians known as Cliff Dwellers told us we were "in the third alphabet." Which charmed Randall, who burst forth happily, "Why, it's like an address in science fiction."

When we told Cal Lowell about this, he, as a former resident, explained that the first, and one-syllable alphabet, was A street, B Street, etc., and sure enough, began at the Capitol. Then came the two-syllable alphabet, "Ashby . . . um . . . Benton . . . Cal . . ."

"Calvert, Cal," Randall chimed.

"And the third alphabet . . . yours . . . starts out Albemarle, Brandywine, Chesapeake? um . . . Davenport. Then, one I call Eliot." A shy smile came and went while Randall knowledgeably corrected, "Ellicot." And the three of us chorused "Fessenden, Garrison, Harrison, Ingomor, JENIFER."

So began our anxiously anticipated Library phase. The jolting and jeopardizing C situation was safely behind us but, alas, had taken its toll. All Randall's fire for his Eliot essay had been extinguished. Permanently.

III

I've always loved Washington and when you live in it, it's best of all.

RANDALL JARRELL
From a letter to poet Elizabeth Bishop

The Washington Zoo

TED RUSSELL

Washington

TO BE TRANSPLANTED from what Randall called "pastoral Greensboro" and our University of North Carolina Woman's College, which he called "Sleeping Beauty," to dwell among the monuments and fountains of the Eisenhower Washington, D.C., that Randall called "the southern city in the north," had only one drawback. No students.

Doing without students during his Army years was a hardship Randall never forgot. When he coined his phrase, "If I were a rich man, I'd pay money to teach," he meant it. He dearly missed students while he was away at the Library of Congress, though he didn't miss grading their papers. To Randall that was always an ordeal requiring several midnight stints before he emerged from his wing chair ashen-faced and hollow-eyed, sighing his familiar "Free. Free. Old Ramble's free at last." And tumble into our bed muttering, "The operation was a success but the doctor died."

Inasmuch as there had not been a Consultant-Laureate since Karl Shapiro several years before, Randall spent the first weeks getting the feel of his official duties and thinking of an agenda. The appointment was *not* funded by taxpayers but by the philanthropy of the Archer Huntington family, with occasional handouts from the Bollingen Foundation and Washingtonian Gertrude Clarke Whittall. The Library donated a suite

and balcony view of the Capitol dome, and poet Ann Chapin
Biddle donated the nineteenth-century furnishings. Phyllis
Armstrong, assistant in poetry, was a tall, tailored, lank-haired
woman who wore wing-tipped Oxfords and whom Randall
admired immensely, half for her expertise and half for resem-
bling T. S. Eliot. Together they pushed out the brittle chairs
and end tables and, I think, a lacquered folding screen to make
room for bookshelves to house the armloads of books Randall
carried up from the labyrinth under the Library. From somewhere Phyllis knew about, they borrowed a long table where
they posed the tall Phaidon and Abrams art books that would
stand alone open at Randall's favorite plates.

One book of enlarged Dürer engravings had about six or
more of St. Jerome, *Der Heilige Hieronymus*, and we would study
them under a reading glass. Sometimes the half-clad saint was
in the desert, kneeling in prayer before his traveling crucifix.
Sometimes he was seated among the rocks with a rock in hand
for breast-beating. Dürer—which Hannah taught Randall to
pronounce Deer-a—always included the lion, and their bond
tenderly reminded Randall of his bond with Kitten, formerly
so often curled asleep on his manuscript pages while he wrote.

We knew many of the painted St. Jeromes: Carpaccio's, Titian's, El Greco's. Randall could spot a Jerome first in any
gallery and we would streak across the room to see it.

A painting that left out the lion left Randall disappointed
and vexed. "Why call him Jerome? Without the lion you can't
tell him from Anthony," he growled. "Bring back that lion!"

The Dürer engraving he liked best was the one of St. Jerome
in his beamed and benched Nurmberger-ish study where the
dog sleeps by the drowsy lion; where the house slippers rest
under the window shelf and a skull sits on top of it. "What a
model of domestic calm," Randall sighed, and I said, "Yes,

there he is. Hunched over his Bible translating away just like someone I know hunched over his *Faust*."

That prompted a mini-lecture (after all, Randall was a *teacher*) on how the dog in the engraving was a catalyst to make that lion indoors believable, and how the skull kept the scene from being "too cozy, like a Dutch genre painting." He pressed my thumb over the skull, "See? See what I mean?" Then, taking it away and smiling happily, he said, "Boy, nothing in art talks like a human skull."

All Randall's empathy with Dürer was equaled if not surpassed by what he felt for Freud. Whether student or friend, you were going to read *The Psychopathology of Everyday Life* and *The Interpretation of Dreams*. "Really," he insisted. "You'll thank me . . . I guarantee." Back in his youthful past was a phenomenal month he'd spent on Sullivan's Island, South Carolina, in the company of Freud's friend and disciple Hans Sachs, which inspired the young psychology major at Vanderbilt to be a psychoanalyst himself until he "had to learn radiophysics and went over into English where I belonged," Randall wrote James Laughlin. Coincidentally, Randall's birthday, May sixth, was the same day as Freud's. And about the time he wrote his poem "Jerome," didn't we make the acquaintance of a Washington *psychiatrist* who drove a Porsche and rooted for the Redskins but was too Jungian for close friendship?

The Jerome of the poem was a dedicated, overworked psychoanalyst who lived near a zoo and often walked there taking tidbits to feed "his" lion. "Jerome," the poem, is ingeniously split into a Conscious, modern-day, Dr. Jerome level and an Unconscious, dream-state, St. Jerome level. Centuries apart as they are, his two Jeromes have affinities. They are both old. They are both intellectuals. They have pet lions. Each has an icon—Gradiva and the Crucifix—and each in his way is contending with the toads and dragons of everyday life.

Dr. Jerome, after listening all day to his patients' "dreams [that] affright his couch" lies down himself, at midnight, on that couch. Falling into the timeless domain of the Unconscious he drifts into a clinically correct Freudian dream, a dream in which he, the listener by day, becomes a whisperer by night. All night. To the night. The dream turns the night into a dragon who craftily inverts the Master's great dictum: *Where Ego was, there Id shall be* to the heretical *Where Ego is, there Id shall be.* As the Baskerville typeface dissolves into italics the omnipotent dream retrieves Dr. Jerome's memory of Dürer's St. Jerome half naked among rocks with his punitive rock in one hand and his translating stylus in the other. The dream then summons a lion who speaks and an angel who won't, and Guilt and Anxiety permeate the scene. The two old Jeromes are nearly one in the nighttime of the dream . . . *but the night is gone.* Morning has come. The Baskerville is back and Dr. Jerome is his true person placidly taking his walk on the shaded paths in the zoo past a lynx and a leopard, to his lion. That zoo was the Washington Zoo and those shaded paths were in Rock Creek Park where *we* walked past the leopard, past the lion to our . . . our lynx.

Jerome

Each day brings its toad, each night its dragon.
Der heilige Hieronymus—his lion is at the zoo—
Listens, listens. At the long, soft, summer day
Dreams affright his couch, the deep boils like a pot.
As the sun sets, the last patient rises,
Says to him, Father; *trembles, turns away.*

Often, to the lion, the saint said, Son.
To the man the saint says—but the man is gone.

Under a plaque of Gradiva, at gloaming,
The old man boils an egg. When he has eaten
He listens a while. The patients have not stopped.
At midnight, he lies down where his patients lay.

All night the old man whispers to the night.
It listens evenly. The great armored paws
Of its forelegs put together in reflection,
It thinks: Where Ego was, there Id shall be.
The world wrestles with it and is changed into it
And after a long time changes it. The dragon
Listens as the old man says, at dawn: I see
—There is an old man, naked, in a desert, by a cliff.
He has set out his books, his hat, his ink, his shears
Among scorpions, toads, the wild beasts of the desert.
I lie beside him—I am a lion.
He kneels listening. He holds in his left hand
The stone with which he beats his breast, and holds
In his right hand, the pen with which he puts
Into his book, the words of the angel:
The angel up into whose face he looks.
But the angel does not speak. *He looks into the face*
Of the night, and the night says—but the night is gone.
He has slept . . . At morning, when man's flesh is young
And man's soul thankful for it knows not what,
The air is washed, and smells of boiling coffee,
And the sun lights it. The old man walks placidly
To the grocer's; walks on, under leaves, in light,
To a lynx, a leopard—he has come:

The man holds out a lump of liver to the lion,
And the lion licks the man's hand with his tongue.

The lynx canadensis *is shy and avoids the vicinity of man,* the sign says on his cage. But somehow an Alaskan expedition captured this one and put him in the zoo. He was usually asleep in his den when we came to visit him on the way home from Randall's work at the Library. I'd bend over the guardrail and call him softly so as not to alert the watchman. "Beautiful? Beautiful? Come see us? We're here. Beautiful?" The lynx would hesitate a moment in his doorway, his silver eyes blinking in the light, and pad over to us with his ear points high and his collar points pointing downward under his chin. He'd sit by us on his side of the cage waiting expectantly. "So intelligent," Randall said, peering into his face. "So like Kitten." We'd find a stick and spear the turkey or liver on it to put through the mesh. The best thing we could bring the lynx was roasted duck skin. He would toss it in the air and pounce on it over and over. Then he crouched beside it to rub it with his cheek and finally gulp it down. Even when the food was gone he would stay outside with us and let us talk to him and take snapshots. "Imagine being liked by a lynx," Randall said.

For that matter imagine being liked by a French waiter at the Colony who told us about racing his kayak in the Potomac. Or . . . or imagine having home-baked beans and brown bread with our Georgetown antiques dealers upstairs over their shop.

"What a joy," Randall often said about the *Washington Post,* especially the stories that he carefully tore out of the paper about families who kept ocelots or chimpanzees or mountain lions. I remember one morning when he said to me over the buckwheats and crystallized Mount Hymettus honey, "Sometime *we* ought to do that, baby. Wouldn't you like to? Live somewhere where we could? *We* could have a mountain lion. They make wonderful pets, you know. They're very gentle.

They don't attack *people*." Domestic fleas didn't either, according to Randall.

"How about a house-broken rabbit?" I said. Mischievously.

And he said, pretending to be gruff, "Von S., I'm going to kiss you."

In this second marriage for each of us we tried to work around each other's inflexibilities as much as possible. Two of mine were no reading at the dinner table and no cats—not even Kitten—in the matrimonial bed. Somehow we never did live anywhere where we could have had a mountain lion.

In Washington Plain Old Pearson's liquor store upgraded Randall from Löwenbrau to Pilsner Urquell and taught him his way around the Rieslings. Brooks Brothers made him taller and Elizabeth Arden made me a blond. Washington catered to all our enthusiasms and we settled in for two incomparable years of hearing the Budapest and their four Strads *live* at the Library instead of recorded on our pitted seventy-eights. Season tickets to the Opera Society in its infancy and the Redskins in their prime. Bonnard, Vuillard, Cézanne, Degas, Eakins, and Matisse? Just down the road at the Phillips Gallery. Vermeer and Donatello just downtown at the National.

In an interview with Mary McGrory for the *Post*, Randall exclaimed, "Oh, Washington would be ideal if I could just find a good singles partner." Hardly was the copy off the press than our phone rang and a fine singles partner named Larry Jaffe volunteered. They divided their tennis time between the Sheraton courts and Larry's Jewish country club, where, one afternoon, bearded Randall was overjoyed at being mistaken for the new rabbi.

And then there was Cumberland, Maryland, where the Conestoga wagons headed out for the Oregon Trail and we headed out for the sports car races. Accommodations were short on race weekends, and we were obliged to consider a clean but humble room for rent in a railroad worker's home. Exchanging eye signals we acknowledged this was not quite our style. However, Randall whispered quietly, "What do we care, best beloved? It's on Goethe Street. Providential!"

After he came under Hannah's sway, Goethe (whom she taught Randall to pronounce Geu-ta), Freud, the Grimms, and Dürer were foundational in a German *gestalt* of Randall's. Mahler, Lehman, Schwarzkopf, and Rilke were part of it. And the early outcome was two companion poems when Randall transformed Dürer's engraving of *The Knight, Death, and the Devil* into a poem of the same name, pairing it with his wry, Faustian dialogue, "A Conversation with the Devil." Gottfried Rosenbaum—Randall's tennis-playing, sports car–driving alter ego in the novel *Pictures*—was already established, and during the Cumberland-Washington days the Rilke translations multiplied. When *The Archangel's Song* was finished the whole translation of *Faust: Part I* began. Add to this the minutiae like Randall's German officer's full-length leather coat, the German wines, the German car, and my maiden name. Of course we stayed in the railwayman's spare room. And gladly. By this time I knew that we, and the Mercedes, would have camped in a sand pile if it had been on Goethe Street.

In reviving the long-vacant position at the Library it was Randall's idea to invite poets to record conversation as well as their work. Among the many chosen were Leonie Adams, Bogan, Bemelmans, Berryman, Frost, Graves, Hoskins, Kohler, Kroeber, Cal Lowell, by all means; and Mauldin, Nabokov, Ransom, Schwartz, Shapiro, and Peter Taylor. Typical of his personally composed letters was one to Elizabeth Bishop when Cal told him she would be visiting in the States before long.

I'm awfully glad you're coming. I'd like to ask a favor, if you can bear to do it. Would you stay (in Washington) until Monday morning and read some poems aloud to me in the process of getting them recorded? I can arrange to have the engineer stay out of the way and never say a thing. I'd like to hear you read them, all the new ones especially, and I'd like to have a recording of them *exist*. Probably both of us will live to be eighty-three. I have a serious feeling about that particular age. BUT the world is full of bombs and airplane crashes and you really ought to read them . . . and I'd like to be able to have them on a record the public could buy.

After Randall's formal introductory Library lecture, "The Taste of the Age," received coverage in the *Washington Post*, *Time*, and *Newsweek*, Randall had calls for poetry readings or lectures from Georgetown University and Howard; from the Congressional Wives and the Congressional Pages; and, among others, from Sidwell Friends' School, the Naval Academy Faculty Wives, and St. Mary's Convent College. Also, there were in-house calls from congressional staffers needing to settle arguments as to "Who wrote 'The Death of the Hired Man'? Whitman? Sandburg? Or?"

Further afield was his invitational address for the National Book Award ceremonies: "About Popular Culture," in which Randall stated, "If it is Mr. Cerf's (chief editor at Random House) id that publishes *Don't Go Near the Water*, it is his superego that publishes Mr. Warren's poems and it is Mr. Warren's superego or muse or daemon, that makes him write poems like 'Brother to Dragons' and 'Promises' when he could be writing best-selling novels. We are safe as long as these men's superegos survive." Continuing in a nervy attack on the publishing business for serving both God and Mammon, Randall referred at length to Ernest van den Haag's book *The Fabric of So-*

ciety with its many sharp twists of the knife, and quoted Santayana's words, "It is worth living in the twentieth century to get to read Proust." Then he asked, "Is it worth it to get to read *Peyton Place*? We ought to say what we know. It's better to read Proust or Frost or Faulkner . . . better in every way: and we ought to do all we can to make it possible for everybody to know this from personal experience. When we make people satisfied to have read *Peyton Place* and satisfied not to have read *Swann's Way* we are enemies of our culture . . . and Jefferson and Franklin and Adams would look at us not with puzzled respect but with disgust and despair."

During his talk the audience fidgeted and whispered and Ayn Rand said audibly, "Ach, thiss man, mein Gott! He should not be allow-ed. Vy don't ve leaf?" At the conclusion there was mild applause and three persons came forward to congratulate Randall: Ernst van den Haag, Ralph Ellison, and James Jones. Dinner followed at Toots Shor's.

Another such invitation came from the Dallas Art Museum for Randall to join Meyer Schapiro on their arts festival program. Schapiro lauded Cézanne, and Randall compared abstract expressionism to the published work of Betsy, the finger-painting chimpanzee at the Baltimore Zoo.

Still another official invitation came from San Francisco, where we spent a month visiting colleges and universities in the Bay Area, and where San Franciscan Kenneth Rexroth took us to a Chinese restaurant for dinner. Actually, Randall particularly enjoyed Chinese food and reading Chinese menus but Rexroth had preordered the meal, including, he said, "really authentic" dishes. Nothing looked familiar, and when the lid was lifted off one unsightly conglomeration, Randall said, "What, pray, is that?" Rexroth discoursed at some length, peppering his speech with Chinese phrases until Randall could stand it no longer and burst out, "And they scoop it out of hollow trees?" Rexroth served himself, saying nothing, and Randall ate fried noodles.

At San Francisco State, where he gave his Wallace Stevens lecture, we met our first Beat, Gregory Corso, who seemed as taken with us as we were with him. He had the appeal of a streetwise waif with, perhaps, the potential of a Villon. With little more than grade school preparation he was so . . . so pleasingly promising that someone made it possible for him to audit classes at Harvard, where he wrote his first book of poems, *The Vestal Lady on Brattle*. Inscribing a copy for us Corso wrote, "For Randall and Mary Jarrell in return for that wonderful, gentle evening in San Francisco."

We had taken the vegetarian Corso to Fisherman's Wharf, where he ordered a fruit salad and impressed us with his hopes for loving the world and living his life for poetry. Later that evening we met with Allen Ginsberg and William Burroughs, who challenged Randall to demonstrate "excellence" in poetry. When he read Frost's "Home Burial" aloud to them, Ginsberg leaped up in the middle of it shouting, "Coupla squares yakking!" Randall finished the poem and was finished with Ginsberg.

Later, Corso left San Francisco en route to New York and stopped off in Washington to see us. He arrived in his Harvard chinos and tennis shoes, wearing a black Shetland sweater over a white Arrow shirt. He was medium height and Mediterranean-looking with a jutting brow, heavy jaw, and thick, black, blown-about hair. He was coatless, tieless, sockless, and penniless—all of which Randall gladly supplied. Then, remembering his own hard times as a beginning poet, and what the hospitality of Tate and Warren had meant, Randall invited Corso to stay with us indefinitely and work on his poetry.

Soon after that Corso's friend, Jack Kerouac, arrived by Greyhound en route to see his mother for Christmas. He wore a knitted wool skullcap pulled down around his ears and was bundled into a lengthy double-breasted vicuna coat that, he told us, Lord Bovril had given him "straight out of his closet." Thirty-ish, fadingly handsome, Kerouac was likable, responsive, and witty; and there was a sweet melancholy about him that made us yearn over his acute alcoholism. He took no food while he was with us but kept a six-pack of beer always within reach, even carrying one in each hand the day we walked to the zoo.

In his book *Desolation Angels*, Kerouac tells his side of this visit, naming Corso "Raphael" and Randall "Random Varnum."

I see the sweet house with the dim lights and ring the doorbell. Raphael answers.

"You shouldn't be here but I'm the one who told you I was here so here you are."

"Well, does Random mind?"

"No, of course not—but he's upstairs asleep with his wife."

"Is there any booze?"

"He has two beautiful grown daughters you'll see tomorrow—it's a real ball, but it's not for you. We'll go to the zoo in his Mercedes-Benz."

"You got pot?"

"Still got some from Mexico."

So we turned on in the big empty piano living room, and in the morning I see the real horror of it all; in fact, I added to the horror by my really importunate presence . . . All I remember is that incredible Raphael and incredible me were really imposing on this gentle and quiet family, the head of which, Varnum, a bearded kindly Jesuit, I guess, bore (up) with a manly aristocratic grace.

There's Random Varnum the great American poet watching the Mud Bowl on TV over his London Literary Supplement . . . He shows me his poems, which are as beautiful as Merton's and as technical as Lowell's . . . I wheedled in the living room writing poems and talking to the youngest daughter, 14, and the oldest, 18, and wondering where the Jack Daniels bourbon of the house was hidden . . . I felt forlorn and lost, even when Raphael and I and Random's wife went to the Zoo and I saw a female monkey giving a male monkey some skull (or as we call it on the lower East Side, Poontang) and I said, "Did you ever see them practicing fellatio?" The woman blushed and Raphael said, "Don't talk like that"—Where'd *they* ever hear the word *fellatio?*

But we had one fine dinner downtown, Washingtonians startled to see the bearded poet, wearing my huge vicuna coat (which I gave Random in exchange for an Air Force fur-collared leather jacket), to see the two pretty daughters with him, the elegant wife, the tousled, bedraggled black haired Raphael carrying a Boito album and a Gabrielli album, and me (in jeans) all coming in to sit at a back table for beer and chicken. In fact, all miraculously piling out of one tiny Mercedes-Benz.

I foresaw a new dreariness in all this literary success. That night I called a cab to take me to the bus station and downed half a bottle of Jack Daniels while waiting, sitting on a kitchen stool sketching the pretty older daughter . . .

Sensitive to Kerouac's loneliness, Alleyne caringly packed him a lunch for the road and, anxious about his poverty, tucked in two of her completed green-stamp books with in-

structions for how to redeem them. Later, gentlemanly Kerouac sent her a note of thanks and wrote in his book, "I still feel
ashamed about that uninvited visit and haven't done such a
thing since and never will."

Corso's visit continued; and he, taken with Beatrice's un-
abashed adoration of Randall, renamed her Little Boswell and
explained why. In the "tiny Mercedes" the two of them occupied a package space behind the red leather seats that we called
"the well." With knees drawn up under their chins, two
medium-wide, medium-tall persons could just barely squeeze
into the well and ride short distances, knees to knees, face to
face, like a pair of bookends. Corso and Little Boswell went
everywhere with us.

For six weeks Corso wrote a poem a day in the languages of
Whitman, Williams, and Cummings, and he celebrated such
subjects as the yak, the needle, the penis, and cosmic love.
Though Randall's long talks with Corso kept him from his
own poetry, he obligingly took on Tate's role as a mentor.
Corso, however, was committed to the Beat principle that
spontaneity was all and refused to revise, preferring to simply
trash the old poem and start a new one. In writing about the
Beats later, Randall said, "Failure to select, exclude, compress,
or aim toward a work of art, makes it impossible for even a tal-
ented beatnik to write a good poem except by accident." Dis-
enchanted with Corso, Randall was relieved when he resumed
his trip to New York.

Just before Christmas Cal, in the first stages of mania, spent
an afternoon in Washington with us.

Rumpled Cal, somewhat swollen above the neckline of his
collar and clothed in his stodgy last-for-a-lifetime-or-two
jacket, trousers, and shoes, was wrinkle-browed above his spectacles and both gentle and stern below. He was like a Back Bay

Beethoven compared with agile Randall, who was groomed for the library, like a young, serious Chekhov. Chekhov, that is, attired in British tweed and twill and a new Macclesfield tie. Aesthetic challengers, intellectual equals, fond opponents, they had a unique *bruderschaft* dating from their Kenyon days and freezing nights as roommates in Mr. Ransom's attic.

Barely seated, they were instantly engaged in their interminable weighing and sifting of poets from the present back to Homer. A game of "Who's First," Shapiro called it. Great spectator sport to observe them disposing of most all the American poets except Lowell and Jarrell.

Cal spoke in long, halting sentences—somewhat overlong that day—with Randall darting in and out, and Cal, unperturbed, perfecting them as he went along with just the definitive word or phrase he wanted. He animated this, as I'd often seen before, with a downward forefinger circling the air which Randall dubbed "Cal stirring his porridge." Randall, as the fervor mounted, responded by working his mouth as if he had a pebble in it, which I dubbed "Randall chewing his raspberry seed."

Christmas came with apple-orange-prune stuffing for baked Long Island duckling and luscious scraps for our lynx. We and the girls reveled in our last *buche de Noël* from *Avignone Frères*.

Alas, in the new year the grapevine was sizzling with the latest details of Cal's madness. Randall, hating both the message and the messengers, brooded and blamed himself for failing Cal. Not that Cal ever felt that way; in fact, Cal would eulogize Randall, when the time came, for his "uncanny clairvoyance for *helping* friends in subtle precarious moments . . . Twice or thrice, I think, he must have thrown me a lifeline."

Late that spring the cherry blossoms burst forth simultaneously with hedge-high azaleas, blessed or cursed wisteria: tree lilacs and tunneled avenues of white dogwood like driving through the corps de ballet in Sylphides. Several huge buckeyes bloomed on Jenifer Street and our garden bench was in a bower of pink weigelia. Our grass had more violets in it than grass and our own mockingbird was gloriously in voice. Many afternoons, after I picked up Randall at the Library, we put the top down on the Mercedes and headed for Howard-Johnson-on-Wisconsin for takeout milkshakes and to ride around in the spring.

"Washington is giving itself to us," Randall said, half sadly.

Yes. And all the while our time with the Library and Larry Jaffe and the lynx was ticking away.

"Let's not talk about it," he suggested.

What he did talk about on the phone to the improving Cal was what a joy it would be to have classes again in the fall.

"Believe me," he told Cal and others, "I'll never go anywhere again without students. Never."

And yet, Washington had not been a bad tradeoff. During the two years there Randall had written three critical essays and another in process, four poems, nine Rilke translations and—as Randall jubilantly told Cal, "I'm two-fifths done with *Faust*."

Back in Greensboro the Mercedes would miss the factory-trained mechanic in Washington and the stereo would miss the able diagnostician at Shrader Sound who had prescribed the four KLH speakers. And we? How we would long for the zoo, for the Redskins, the Budapest Quartet, and for the hundreds of glorious paintings. But there were consolations. Chancellor Graham had resigned. A full professorship was waiting for Randall as well as his good singles partner, Parole Officer Carrigan. *And the students.*

While we were gone the college for women we had left behind was maturing into university status. Men students would enroll. Black students. Graduate students. And there was a new course Randall had delightedly devised called "The Narrative," which was a survey of taste-elevating literary novels, stories, and poetry. "The Narrative" would introduce the students to a variety of examples, such as Fitzgerald's *Great Gatsby,* Frost's "The Witch of Coös," Kafka's "A Country Doctor," Taylor's "What You Hear from 'Em?" plus Rilke, Forster, Dinesen, Bowen, and beyond, back to Wordsworth's "The Ruined Cottage," Dostoevsky's *Crime and Punishment*, Chekhov's "Rothschild's Fiddle," the book of Jonah, and more. "The Narrative" would be taught in the university *library* lecture hall and to a class of three hundred accompanied by their three hundred exams, their three hundred term papers and final grades, and their three hundred fall semester head colds.

IV

Faust is unique. In one sense, there is nothing like it; and in another sense, everything that has come after it is like it. Spengler called Western man Faustian man, and he was right. If our world should need a tombstone, we'll be able to put on it: HERE LIES DOCTOR FAUST.

RANDALL JARRELL

From the Preface to the book

Translating Faust

SPECIAL COLLECTIONS, JACKSON LIBRARY,
UNIVERSITY OF NORTH CAROLINA AT GREENSBORO

Faust: Part I

THERE ARE TWO QUESTIONS regularly asked about Randall's translation of *Faust,* and one of them is: Why did he do it? Poets don't ask this. Poets know that when you can't write your own poetry you translate someone else's. And, alarmed as the poet is at the temporary disappearance of his art, he can sometimes quiet his psyche by letting another's art—Goethe's or Racine's or Molière's or Homer's—flow through him on the chance that his own will come back.

In the early 1950s, Randall had finished his first book of criticism, *Poetry and the Age*, and *Pictures*, his novel, and he was having a hard time writing poetry. When he got stuck in a poem, he'd sometimes joke about it, saying, "Help! Help! A wicked fairy has turned me into a prose writer!" But it was a black joke. Though he had done a few poems, he'd done more than a few Rilke translations, and in 1957 he turned to translating Goethe's *Faust: Part I.*

Randall's job at the library, mainly, was to be in his office four hours a day to receive anybody—from Kankakee to Lithuania—who was interested in poetry. The Consultant-Laureate was given a thick personal appointment calendar book. No poet could have kept all the appointments there was room for in that book. On the other hand, there *weren't* many people—except from Kankakee and Lithuania—who were interested in meeting the Poetry Consultant-Laureate. One day Randall appropriated this year of empty pages, wrote on its inside cover Goethe's motto. "*Ohne rast aber ohne hast*"—"Without rest but without haste"—and started his translation.

The "Dedication" didn't take long, and he skipped the "Prelude at the Theater" for five years, and he put in his immediate time on the "Archangels' Song" and the first soliloquy in Faust's study. For my birthday, in May, I received these in a little booklet he'd stapled together and illustrated. One drawing was of the mists of Heaven and the flames of Hell with Earth floating between; above the planet and shining on it, he'd put our symbol, two stars in a crescent moon. The other drawing had G-O-E-T-H-E and F-A-U-S-T in wavering letters with a detailed sketch of a bearded, aquiline-featured face much like Randall's own.

At first, when people asked him why he was translating *Faust*, he told them, "Oh, it's something I've been meaning to do." And that was true. Later on, when the question came up, often from the same people, he'd say with a look of helpless defiance, "Why, for Goethe!" And that was true, too. It was not done for riches, not for fame, not for a foundation grant; it was an assignment of the soul. In a taped interview about this he commented: "At least, if I can't work on poetry of my own, I'm working on poetry better than my own . . ." And the real value of this to a poet is that it fixes him in the realm of poetry, where he can still regard himself—though poemless—as a *poet* translating a *poet*.

Having written poetry, Randall found the realm of prose no substitute; if anything, it was a threat. The afternoon we first met, he was soon telling me that a poet is never certain about being one. I didn't understand. Randall was certainly a poet to me. But he insisted that the best poems often came in one's youth, like the breakthroughs in science, and after that were likely to diminish. He cited Frost's middle years and William Carlos Williams's. "And there's a good chance your poems will dry up altogether," he said. "It's a risky business."

I often saw that the little control Randall had over his art— or what he thought of as the little control he had—haunted him. Between books he could even imagine that his last book might be his last. Translating *Faust* insulated him from the worst of these fears, and doing so "for Goethe" exalted his efforts. Of course he shelved it when there was a chance of writing a poem of his own, but each time he came back to it he was as determined as ever to complete it without rest but without haste, for Randall believed in the worth of *Faust* for our age.

And does *Faust* interest readers and actors today? It hasn't so far. But *Faust* in Germany, a hundred and fifty years old, is read, performed, and worshipped. Here it has few readers, no performers, and one worshipper—"a bored anachronism," as Randall labeled himself in his poem "A Conversation with the Devil."

In that poem he signs no contract, but in this translation he made a pact with himself to bring *Faust* across the ocean and across the years with its humanness intact. For it is Goethe's emotional investment in these archetypes that keeps them going in his own country. When *Faust* travels to America, Goethe's fully felt emotions are hidden in the character's linings, as if to declare Faust's and Gretchen's woe and despair and rapture and bliss were illegal. Randall's translation declares them. Parts of *Faust* that have seemed dull and neoclassical to others' sensibilities were to Randall sharp and relevant: sweet and true; raw and live. And the tone and language he used are neither imitation German nor imitation English, but plain English intended for readers and playgoers.

The other question most frequently asked about this *Faust* is: How was it done?

Well, he did not start at page one and go through to the end. Randall worked on it spasmodically, a whole scene for a week or two, and then "The Song of the Flea"; some days the dark monologues, some nights the bright dialogues. When he felt Faustian, or Mephistophelian, he worked on their lines to the exclusion of the others, the way Grandma Moses added green to a series of paintings she'd lined up and felt like adding green to.

In translating he was as much at the mercy of his moods as he was in creating; but it seemed, to him, right to be. Randall held that the poet-translator was not an able bilingual who could compete with a computer. Or else, that *was* what he was if he was uninvolved with the work of art, that is, if he kept him*self* out of it, barred him*self* to it. He worked in the spirit of Goethe's words: "If the translator really understands his author, he can evoke in his own mind not only what the author has done, but also what he wanted and ought to have done. That at least is the line I have always taken in translation . . ."

To bring this off, the translator and his poet must start out as friends, grow into associates, and at instances—beyond command—become one flesh almost. When that happens for a few hours, or a few days, the translator's persona recedes, is out of the way. Then fresh words spill over him not weighed or thought of before. New ways of using them freshen what has staled. Handwriting spreads, pages fill, and words come that are inimitable and stay in the text forever.

When this happened Randall came to lunch exhausted but exuberant and would tell me in a marveling voice: "Was I lucky! It's been going like a house afire!" When the opposite of this happened he'd quote Faust's "Woe follows joy! Joy follows woe!" and labor on. If it got too toilsome, though, he'd stop and we'd drive to the tennis courts, or if it were summer, to the beach.

Between the heights, when the translation was God-given, and the depths, when it was impossible, stretched the fertile plain of Randall's abiding empathy for Goethe. Through all the difficulties that that idiosyncratic poet made for his translator, Randall was heartened time and again by the advantage he had—for once—in being, as Goethe was, a Romantic. Between them they had the congeniality a poet and his translator must have for the better part of their journey together or they weary of each other and part. Goethe said, "Fifteen minutes and we tire of any landscape," but in eight years Randall never tired of Goethe.

Randall's hero, whom Napoleon called "the other genius," had been a brilliant university student, superior to his professors, but the equal—by his own statement—of his friend Herder, the theologian. At twenty-two Goethe was practicing law and attracting the attention of his colleague's fiancée. At twenty-five he published *The Sorrows of Young Werther* and got the attention of all Europe. Still in his twenties, he was advising the Duke of Weimar on mining, agriculture, war, and law; and by the time he was thirty-two he was Minister of Finance and the most powerful man at the Weimar court after the duke.

Goethe lived beyond eighty, continuing over the decades to influence the cultivated world with his plays, novels, essays, ballads, and poems. Somehow there was time for him to direct the theater at Weimar; to study botany, geology, zoology, Greek art, philosophy, and meteorology; to discover the intermaxillary bone in humans; to write a treatise, *The Science of Color*, that is still a standard work in optics; to participate in the allied campaign against the French Revolution, and to travel extensively; to write journals, an autobiography, and thousands of letters.

Goethe also found time for many friends (among the closest was Schiller) and many *amours*. One of the earliest of these, Friedericke Brion, was a provincial and religious girl whom he tired of and left, but whom some view as a part model for Gretchen. Another attachment was Frau von Stein, intelligent, older, and married, whom Goethe saw or wrote to every day for seven years. When that ended, he was nearly forty and he took Christiane Vulpius, twenty, into his home and in later years married her. After her death, he was in his sixties and fell in love with Marianne von Willemer, who inspired the Divan poems. In his seventies—still attracting women—Goethe fell in love once more—with the eighteen-year-old Ulrike von Leverzow.

This Goethe who was considered so great in his lifetime that the biographers began before he died and haven't stopped; this Goethe who elicited that remark from Napoleon, "*Voilà, c'est un homme!*," appealed to Randall to the point of devotion, and influenced him to the point of including *himself* in that phrase current among Germans, "We are the people of Goethe."

As for actually speaking German, Randall didn't—but mostly because he wouldn't. In Germany he could order hen soup and blue trout, interpret sung Mahler for me, and sight-read the ever-present directions, regulations, and warnings so dear to the Teutonic temperament. He could have conversed—how I conversed knowing nothing!—but he was shy of sounding ridiculous. Touring the Werthersee with my relatives, he preferred to bask quietly in the beauty of their umlauts with an occasional affirming *Gut* or *Schön* or *Wir Verstehen*—but then astonish them by looking out the window and exclaiming from *Die Winterreise*: "*Ach, das die Luft so ruhig,/Ach, das die Welt so licht!"* ("Ah, that the air is so calm/Ah, and the world so bright!")

And with *that* he translated *Faust*? No. But it must be said that a first-year student of German can make a comprehensible "translation" of Part I. Really! After about eight months, a student can render Goethe's gold into a useful kind of straw. That is not hard. The trick is to turn it into gold again, and that is where the time went in this translation.

Randall was forty-three when he took on *Faust* and old enough to know some of the geography of his art. Just as he knew his serve was the weakest part of his strong tennis game, he knew his lyricism would be the essential strength in translating Goethe and *that*, with being a natural dramatic monologist, would make a good combination.

As a poet who had read *Faust* since high-school days, Randall knew Goethe's idioms and vocabulary and styles as if they were his own—as, indeed, some of them were. And as one who lived by teaching and writing poetry he knew what he needed to know about meters and rhythms—even Goethe's improvisations of them. His highest aim was "to make of Goethe's poetry in German—where I can—Goethe's poetry in English."

Except for the prose in Part I, Randall translated in metered verse. He did not choose to match rhyme for rhyme because he said "rhyme for rhyme's sake" frequently kills the poetry, and twentieth-century ears are prose-attuned and often find steady rhyming obtrusive. Randall said cheerfully, "My motto is: Anything that wants to rhyme can." But he wouldn't force it.

While the act of translation is direct communication, the *art* of translation is indirect and intuitional. We start with the premise that the original can only be "the original" in the original. And the instant it is approached with another language in mind a like, or unlike, shadow falls on it, and we, the beholders, are enraged or enrapt at what follows. Concerning the work of art Randall wanted to build on Goethe's work of art, the essence was words. As Faust says while he translates St. John, "In the beginning was the Word," and in the end, the translator's prayer is for the word—and tomorrow another, and after that, another, and another—that has the right meaning, the right syllables, the right stress, and, Lord, the right glow.

Aesthetically Randall was on his own, but technically he had the usual ponies with German/English side by side, others' translations, some *Deutsches Wörterbücher*, and Roget. Sometimes he worked out snatches of lines in English on the pages of these books. Then he eagerly copied however few there were onto the unlined paper he liked to use, setting them into verse "to see how it looked," and where the words or lines were missing he supplied the tiny curves and batons of its meter. At the side of his sheets, even the typed ones, Randall often made columns of these little hieroglyphs that totaled an iambic pentameter line and could act as an abacus when he counted poetic feet in place of the missing words.

He overcame most of these as the years went by, but a handful were not resolved to his satisfaction at the time of his death. They were mostly in the Walpurgis Night's Dream Scene, and those lines have been filled in from the handwritten alternate versions that Randall left among his papers. The only complete scene he had not done, and fortunately it is short, was Gretchen's "Spinning Song," which has been posthumously rendered by Cal.

In accord with Goethe's words—that the translator evoke in his own mind what the author has done, wanted to do, and ought to have done—Randall drew on emotions of his own to match Goethe's wherever he could. In speaking of this, he said to me that any success there is in translating, and all the joy, comes when the translator "sees through the other's eyes." And he had little trouble seeing through Faust's.

That both Faust and Randall were intellectuals and professors was a happy coincidence, but their likeness in philosophy was almost uncanny. To open Randall's *Complete Poems* at any page is to find in some degree a Faustian world of disappointment or self-disappointment, and it is to look in vain for that moment so fair that he'd say to it, "Stay!" The yearning tone in the poems, their hope for something more, something different, something new, seems an echo once removed from that craving of Faust, who, bored and desperate, has come to the end of law, medicine, philosophy, theology, and magic—and nearly to the end of his life as he contemplates a cup of poison.

A poem of his own that came to Randall during the translation of Faust's most despondent moods was "The Woman at the Washington Zoo." It is provocative, in a way, to read it with Faust in mind: Faust, caged in his gloomy study with his books and beakers, unloved and unloving, crying out to the moonlight lines that take longer but are not unlike the woman's "change me, change me!" The pathetic ridiculousness of Faust's late lust seems epitomized in another poem, "Gleaning," where "A girl, a grownup, giggling, grey-haired girl—/Gasps: 'More, more!'"

Again, is it Faust or the translator of *Faust* who complains: ". . . and yet, the ways we miss our lives are life./Yet . . . yet . . . /to have one's life add up to *yet*!"? And finally, Faust's longing-and-guilt. Where Goethe saw it as Everyman's, Randall saw it as Randall's. And as a poet who knew the taste and smell and shape of longing-and-guilt in war, women, children, and dreams, Randall was at home with them in translating *Faust*.

About Mephistopheles, Randall said, "I could do his lines in my sleep!" Friends and foes of Randall's know exactly what that meant: know, that is, of the streak in him that could handle with wit and bite and crushing elegance anything a devil would come up with.

As for Wagner, Faust's assistant, Randall shared Goethe's irony for the blinkered academician who sees the obvious, is blind to the significant, and is hopelessly ignorant of the difference.

Martha was the word made flesh from Randall's line in the poem "Woman": "*Are you as mercenary as the surveys show?*" Randall relished her transparently conniving nature, just as Mephistopheles did, and toyed with it with the same hearty, negative enthusiasm.

Valentine was quite the reverse. His character was so vain, boastful, vulgar, and unfeeling, such a nexus of dislikable traits, that he was thoroughly distasteful to Randall. "The sooner he's killed in that duel, the happier I'm going to be!" Randall said. And an odd incident arose out of this. Randall had given me a canary for Valentine's Day and we were chatting about what to name him. Because of the occasion, and because Randall was, as he put it, "a sucker for anything in the twenties," I said, "How about naming him Valentino?" Immediately, Randall's face crumpled and he said, "No! No! Let's not call him *that*!" Ultimately we agreed on New Moon for his name. Still, I was perplexed and had to ask Randall what had upset him about the other. He looked troubled all over again and said, "Oh—oh—because of Valentine in *Faust*, I guess." And he shook his head and added, "Oh, I'm so silly!"

The scene he liked the best, and which was the easiest for him—and the hardest for Goethe—was the Dungeon Scene. This is not to claim that Randall was so right for it (though he was) but to say that it is so right in itself. All that precedes it for three hours—philosophy, blasphemy, sorcery, seduction, murder, and orgy—all are surpassed in the quarter-hour love-death dialogue between the guilt-driven Faust and the half-mad Gretchen. Though we know that Faust has done all that he can to earn his passage to Hell and that Gretchen, at last, will renounce him, *we* don't renounce him. For the mature Goethe imbued both characters with such rich emotions that Faust's torment torments *us*—abandoners ourselves—so that we ache with each plea he makes to rescue her and are pained with each new proof that it is too late. While we are meant to be gladdened by Gretchen's redemption, Goethe gives her such lines in her madness that we grieve for the Gretchen in ourselves who, despite betrayal, recalls as she does the rapture love can bring.

The Dungeon Scene triumphs over the crudest translation or performance and, read aloud, simply by itself with a brief synopsis, can be a moving experience for audiences. We knew this because as soon as Randall completed it he began reading it aloud when he was asked to give readings. The first time was at St. Mary's Convent College in Washington, and at the end the audience was too overcome to applaud or stir and took several seconds to come back to reality. When we left, we too were somewhat dazed by the effect of the Dungeon Scene, and had not come back to reality either. At the car, Randall laid the binder containing the whole thick *Faust* translation on the top of the Mercedes while he unlocked the door. Then we drove across town to Chevy Chase, had dinner, and were listening to the classical music station when we discovered *Faust* was missing.

With hardly a word between us we got in the car and sped through nighttime Washington, sick at heart that all Randall's work could be in some trash can, God knew where, and we'd never see it again. As we drove near the fenced grounds of the

Convent, our headlights glared on sheets of white paper blown against the chain links, dozens of them, and I cried out, "There they are! There they *are*!" In an instant we were out of the car and running to the fence, gathering them up to our breasts, laughing with relief and then laughing through tears of joy. Some had been walked over on the sidewalk. Others were spattered in the gutter. One, in the street, had the print of tire marks on it. We even found the binder, whole and unhurt. And back at home we stayed up until the small hours jubilantly smoothing and putting in order all the dear pages and clamping them inside their cover again.

When Randall returned to teaching in Greensboro, the translation was what he called "three-fifths done." (Randall liked to divide by fifths.) The classes took more of his time than the Library of Congress had, but he went back to Faust each summer. In Montecito he finished the long put-off "Prelude at the Theater," working on it each morning in a patio in the shade of a two-story yellow Banksia rose vine. In Italy Faust made progress again—just as it had when Goethe was in Italy; but by this time three fairly long poems had interrupted the translation and Randall and Atheneum, his publisher, put these with the ones he'd had on hand to make half of his book *The Woman at the Washington Zoo*. The other half, regrettably to him, were Rilke translations. Not that he didn't love Rilke and love translating him, but this book, only half his own poems, made him feel only half a poet.

That fall he retranslated a particularly despairing monologue of Faust's as he saw through his eyes more clearly than ever the cares the world inflicts and the rewards it holds back.

> *You tremble for all that doesn't happen,*
> *You weep for everything you've never lost.*

In 1964 *Faust* was all but finished. Several sections of it had appeared in print and only that handful of lines and the "Spinning Song" were eluding him. Intermittently, as a sort of refuge in a vacant hour, Randall would type and retype pages. He'd comma and un-comma lines and he'd neatly block out a word only to neatly reprint it. His prose was flourishing again. The second book of criticism, *A Sad Heart at the Supermarket*, came out, and four children's books were written, one after the other.

In the last year of his life, Randall's *Faust* books, the hundreds of worksheets, and the broken Wildhagen were stored together in a pie safe we'd just bought. It was in a room where he was spending more and more time. He kept the Roget out and was now putting it to use with the *terza rima* of a new, lengthy, autobiographical poem he was suddenly caught up in. The more he wrote of this poem, the more came to him, and the more the habit of poetry returned and fastened on him again. Mysteriously and thrillingly he worked at it day and night. Prose and *Faust* were not to be thought of, but everything he thought of made poems and made his final book, *The Lost World*, which Cal said was his best work.

How gaily he announced to audiences that he had *new* poems to read to them and that it felt "wonderful, the way Goethe felt when he said, 'You know, I'm a born writer.'" Randall would add, "And I almost *never* feel like a born writer."

All this took place in the same years that Lowell was translating Racine's *Phaedra*, Richard Wilbur Molière's *Tartuffe*, and Robert Fitzgerald *The Odyssey*, and all enjoyed a return of their own poetry. So, yes, I believe this translation Randall did "for Goethe" did the same for him. There, at his side, I saw in the bewildering absence of his own art how an artist endures a time of self-extinction and longs for something to fill that void.

Always the world—like Mephistopheles—is ready with ru-inous offers. Goethe himself had to deal with this, and he did, as is apparent in a paradox of his that Randall delighted in quoting to students and audiences and that I will conclude by quoting for everyone:

"The safest way to avoid the world is through art; and the safest way to be linked to the world is through art."

V

Randall was most awesomely honest. His highest regard was for the "disinterested" person who had swept away prejudice and self-deceit, and he seemed to me to have succeeded in being that person.

His honesty was related to his generosity to students, and me, and, I think, to people like Christina Stead and Corbière. We were likely to do nothing for him in return, as people not on committees, not writing grant recommendations, not mentioning things to powerful people, not promoting, or appearing. He battled for the survival of such people.

He never modified his opinion of a book for a political or expedient reason, though he sometimes changed his mind. He called names in his reviews, the names of the obtuse and the vicious. He had bitter enemies in the casualties of his honesty, but he never seemed to feel this price too much for self-honor, though I think he suffered sometimes, in rare moments, at certain worldly honors appropriate to him he never got. The important thing was that your work was good and that you did it. We all believe this, but who else lives by it?

ELEANOR ROSS TAYLOR
From *Randall Jarrell: 1914–1965*

Randall and Peter: Greensboro

SPECIAL COLLECTIONS, JACKSON LIBRARY,
UNIVERSITY OF NORTH CAROLINA AT GREENSBORO

Peter and Randall

LIKE CHEKHOV, PETER TAYLOR was an element of Randall's makeup with whom you were indoctrinated, and if the name of "the best fiction writer around" wasn't as familiar to you as the other you got an astonished stare from Randall and felt dismissed as tasteless and ill-read.

They met in the thirties at Vanderbilt in the aura of Allen and Red and Mr. Ransom. And they became *real* friends after the war when they were two literate Tennesseans in the wilds of New York. Before my time, they'd summered at Cape Cod and weathered a winter or two in a duplex in Greensboro where nobody mowed the lawn but Peter, where a Mahler symphony on the Jarrell side came through the walls to the Taylor side with perfect fidelity, but also where Randall praised and criticized and conferred invaluably with Peter about *A Woman of Means*, some of his stories, and literature in general.

In the fourteen years after I met Peter he owned nine houses to our one, and rented at least six villas and a flat in Europe to our one. At the same time he'd seriously inquired about, and nearly got, twice that number; and read ads for, driven by, and fantasized over twice as many more. Peter was a House Man. Randall was a Car Man. And Eleanor Taylor and I followed their bents. The Jarrells thought the Taylors were a kind of "Main Chance" for aging houses. There they were, we said, two *hot* writers who kept taking on some old house plagued with sinking landfill or leaking roofs or both; and making it well, releasing it, and taking on another.

On the flyleaf of our copy of *Miss Leonora When Last Seen* Peter wrote: *With love from the Fisher Park Taylors and especially the head of the house (meaning the one who likes houses best).* "Exploring" the countryside, Peter could see a onetime *mar*-velous *end*-chimney house or center-hall house in every forlorn old farm dwelling stuffed with hay.

When Randall talked literature, composers, or painters Peter listened and absorbed. When Peter talked St. Louis, his grand*faw*ther, or parties we listened and thousands of *New Yorker* dollars went out the window. When Peter wasn't writing or hanging wallpaper it seemed to us he craved parties as much as Randall detested them. When it was one of the Taylor's parties, we certainly went, but Randall would grumble, "Peter likes too many people." We liked too few. At the party, Randall—worn and dehydrated from a blazing afternoon on the tennis court—collapsed on the sofa, consumed a gallon of orange juice, and contributed as little as possible socially. On the other hand, Peter—who, like as not, had spent the afternoon laying the brick walk up to his front steps—looked showered, flanneled, and radiant. In love with the party, and hoping it would last all night.

"Give me parties by the aesthetic method," was what Randall said when referring to parties recollected in moments of tranquillity and when we had Peter all to ourselves. This was often in the evening when we were rocking on the dark veranda of "the old Office" in Monteagle, or perhaps in the porch swing behind the wisteria vine at Fisher Park. Time flew as Peter presented Randall—sparing him the strain of actual encounter—with the highlights from parties with the Bishop of

Sewanee, or the dowager Mrs. Polk from the Delta, or Irene Castle, whom he reported was raising Great Pyrenees.

Peter's party stories usually had a sprinkling of what my St. Louis mother called "prominent people" (compared with "nobodies"), and Peter's enchantment with these was to Randall both appealing and naive. Peter's account of his thrilling discovery at the Royal Court Theatre that he was seated directly back of Princess Margaret brought an eons-older smile to Randall's face. Never mind, I thought, Princess Margaret would have thrilled me, too; and it was all due to this "weakness" of Peter's that we wound up that summer in Levanto sipping Negronis on the Louis Seize furniture of the hand-kissing Baron Massola.

Except for Peter the Jarrells would have stuck in their Cape Cod rut forever, I think, but one spring in Columbus changed all that. The Taylors were living in their twenties Spanish with-the-powder-room-under-the-stairs and the Jarrells were on one of those hectic, overprogrammed, university-paid lecture trips that Peter and Randall wangled in order to see each other. These were better than nothing, we all agreed, but Randall called them "scrappy visits that hardly count." Between cafeteria lunches and cutting across parking lots we'd catch up on Cal, Allen, Jean Stafford, Red, Cleanth Brooks, Katherine Anne Porter, Robie McCaulay, the Ransoms, et al., but there was too little time for Henry James or Tolstoy or Proust.

On our last morning there—I seem to remember having waffles and coffee in the breakfast nook—Peter had just said, "I know people should never bore other people with their dreams," and told us one (which certainly didn't bore old dream analyzers like us for one second) and then suddenly remembered the important news that the Taylors had rented a villa with French doors and fruit trees in Buonassola, and why didn't we come, too? Before we could say, "Yes," "No," or "Where's that?" Peter produced from that quadrant of his brain that controlled "Ril Estate" a roof garden flat for us in nearby Levanto, with a little English library, a half-day maid, and a dressing *cabina* at the beach with a private plot of sand and "*umbrella*." We took it.

That summer Peter's and Randall's friendship expanded its capacity to seven. There were Peter/Eleanor and Randall/Mary; their Katie and my Beatrice; and also their Petey, "our" youngest member, who called time on us to eat, to go home, and to talk pterodactyls with him or the Arch of Titus. Every time Peter softly cautioned him, "Now, Petey, don't be a pest," it evoked from my childhood my mother's, "Now, Mary, don't make a scene."

Because of Petey we rarely met at night, and because of the writers we rarely met in the morning; but we often met for picnics and swims and tea at the *pasticceria*. We explored La Spezia, where Shelley drowned; Chiavari, where those spindly gilded chairs Jackie Kennedy used in the White House are made; and Rapallo, where the Taylors had once rented a house.

In Rapallo the little girls longed to see the sights by white-fringed *carozza,* which Randall thought r-dick-a-lus and some of the Taylors feared was too much for the horse, but the driver won out. No sooner had we wedged and layered our seven selves into the *carozza* than we realized Rapallo was all hills. At once, those who'd felt r-dick-a-lus felt increasingly so and sat in uneasy silence while the horse labored upward, ever upward to The House Where the Taylors Stayed. Then, on the downward sprint, those who'd feared for the horse feared even more for themselves and clutched leather staring starkly straight ahead, all unmindful of the great estates whizzing past where Peter said Pound and Lawrence and Beerbohm had stayed.

It was a summer of Chekhovian vignettes. In the *ristorante*, all of us starving and impatient except Peter, patiently diverting Petey with a car-chase story concluding with the quote "and they all came to grief on the old Rocky Road," complete with cartoons on the paper napkins. In the little English library, Beatrice on her couch looking up from *Major Barbara* and asking Katie on her couch what was happening in her book. And Katie answering, "Um . . . well . . . oh . . . Christian and Hopeful have just gotten to Doubting Castle and are talking to Giant Despair." And Peter and I sunning ourselves on the little mole looking toward the beach at bearded Randall, browned to the waist, animatedly going over the final manuscript of Eleanor's poems for *Wilderness of Ladies*. Like characters in a silent movie Randall read, head down, and then exclaimed, head up, and then bent toward his colleague admiringly, and talked to her in a fiercely determined way: Eleanor, knees under, and speechless, gazed out to sea smiling her half-believing, half-embarrassed smile.

Meanwhile back at the mole, Peter and I—trying tactfully to stay away a little longer—had completely run out of conversation and were mindlessly talking about Eloise Hoblitzelle, whom I had never met but shared a legacy with and whom Peter had never met but thought he knew of because "Hoblitzelle is a good name in St. Louis." I'd begun to think Peter must feel as if he were stuck with me at a dance and it was a welcome sight when Randall stood up on the sand and waved us in. How we all blended that summer like one household at two addresses. How bereft the Jarrells were at the end with tears in their eyes when the Taylors set off for Rome with stars in their eyes and a dozen new abodes ahead to contemplate.

Our copy of *Happy Families Are All Alike* is inscribed: *To Randall and Mary and Beatrice with love from all of us—and let's start making plans for next summer.* Peter made them for Monteagle and found us Rhodes' End that summer, and a house near the tennis court the next.

Another summer we went to Antioch on his account. It was a little out of our style but when the summer school dean told us Peter would be there Randall accepted. Actually, Antioch wasn't quite in their style either. In the first place, Peter and Randall "made it a rule" not to teach in the summers. In the second place, their family of four, and ours of three (plus cats), was not accustomed to graduate student housing. But the final flaw, and perhaps the worst, was that the only place to eat in Antioch was The Lemon Drop with its Dieter's Plate instead of some sway-backed clapboard home on a dirt road—which Peter always knew about in *Tenn*-essee—where a pair of widowed sisters served *mar*-velous meals. As time wore on we wondered more and more why in the world Peter had wanted to come to this place, and when the mood seemed right we asked. Peter looked at us in amazement and cried out, "We? We only came because the summer school dean said *you* were going to be here." When he wrote in *A Long Fourth* for us he said: *For Randall and Mary with love from Peter and Eleanor on the night of August 28 at Antioch!* with Antioch underlined three times.

My mother used to say, "Peter has such a way about him" meaning what St. Louis used to call "an engaging manner" and what the Jarrells called "being irresistible." We knew that inside *our* Peter Taylor was the solitudinous writer, the Tolstoy reader, and the paterfamilias even though on the outside he seemed all youth and charm and the darling of his "prominent family." People liked that. People would open up for Peter and tell all, or as Randall said, "eat out of his hand." He could talk their language, and we'd hear him threatening, as they might, to "go on a diet," or "give up *cig*-arettes," or "write a letter to the newspaper," except when *he* talked that way it was interesting.

For instance, Peter believed his body had a "good" side and a "bad" side, as if you drew an imaginary vertical line down the middle of him, each half would be different. Well, so does everybody's. But Peter's were different—he claimed—because the "good" side was always perfectly okay and the "bad" side always got the sinus trouble, the earache, and the sprained ankle.

Another opinion Peter had that Randall disagreed with was that a wolf could mate with a dog. Peter was always coming up with some story about "a dog that was part wolf" or was it part bear? And once or twice Randall wanted to take exception to argue it out, but in time he let Peter get away with it and would only shake his head and say, "Peter Bell. You're a scream."

Part of the reason he was such good company on trips was that he was never "indifferent" and somehow maintained a kind of sparkling *interest* in all the little life around us. We'd hear him telling Katie he'd have to go into debt again for her allowance, but he'd pay it all back the very minute the *New Yorker* check came and for her to keep track of it. And then he'd go on with such genuine *interest*, "You are keeping track of it, aren't you, Katie? How much do I owe to date?" And when Katie asked, "Mmm . . . Oh . . . Do you mean in lire, Daddy?" Peter said, "Oh, heavens no! Tell me in dollars, Katie, then it won't be so much."

Once in Monteagle in the car when all of us were bound for a picnic at Rutledge Falls or The Old Stone Fort, Randall and I razzed Peter about some signs on trees along the road that said, Pop Taylor's EATS, or Taylor and Taylor Wreckers, or Taylor and Son Cesspools, and Peter said, smiling, "Yes, those are *all* relatives of mine." And when Randall and I laughed and laughed, he said, not laughing at all, "But they rilly are. All the Taylors in *Tenn*-essee *are* related. All kin."

With my mother, Peter again would seem especially interested as I'd watch him talk and listen to her in the lamplight having what he'd later say was a "fascinating conversation" about Mary Institute and porte-cochéres and the Veiled Prophet's Ball. All the while her seventy years would slip away until my mother—like Proust's grandmother—had the face of a girl again. In a book he gave to her, Peter wrote: "*To our very complicated relationship—that is, to my favorite mother of a friend and/or to the mother of my favorite friends.*"

Now, as I finish, I wonder—as I never did at the time—why this friendship worked? Literature is undoubtedly what brought them together. But after that, they were on their own. Why *were* they so close? They weren't much alike. Peter was a romantic and Randall was a romantic intellectual. Peter loved England and drank whiskey; Randall loved Germany and drank Riesling. Peter had children and Randall had cats. Peter quoted people and played bridge. Randall quoted literature and played tennis. Peter wrote fiction and Randall wrote poetry.

Of course that last distinction may have helped the friendship on both sides. To be perfectly honest, I doubt if Randall would have felt quite so cozy if the *New Yorker* were continuously bringing out Peter's *poetry*. Another safeguard, perhaps, was that the temperature of their friendship was like Goldilocks's porridge—not too hot, not too cold, but just right. Nobody was a pest and nobody made a scene, and as Randall wrote my mother that summer, "We're having a dovey time."

VI

Like Randall Jarrell's Bats, we live by hearing, *by* vibra-tions; *by having* heard *what makes us happy—his way of saying what he says.* I cannot think of anyone who gives me *more incentive than Randall Jarrell, as I read him or think about him.*

MARIANNE MOORE

From *Randall Jarrell 1914–1916*

"*My Poetry is Meant to be Heard! It's not just words on a piece of paper.*"

TED RUSSELL

The Lyric Ear

From early to late, Randall's poetry *squeaks, creaks, flaps, whirs,* and *whinnies* like a Stanislavsky script. He wrote poems in which the sea *coos,* thunder *mutters,* and a house *hums,* "We are home, We are home." In another poem a bird calls out, "Red clay, Red clay" or else he's saying, "Directly. Directly." To me a most dazzling feat is the range of tones Ear *heard,* and that Randall described, from a Southern girl's shout of greeting to a friend on the university tennis court next to his:

> . . . *half squeal, half shriek, your laugh of greeting—*
> Then, decrescendo, *bars of that strange speech*
> *In which each sound sets out to seek each other,*
> *Murders its own father, marries its own mother,*
> *And ends as one grand transcendental vowel.*
>
> *"A Girl in a Library," The Complete Poems*

And yet, with the Campbell clan in Nashville, where he was born, or with the Jarrells who had gone West to California, Randall dwelt among talkers of just such pitch and intensity. Had, indeed, spoken "Southern" himself, "Until I made up my mind not to."

"And when was that?"

"Oh, I don't know, peaches. Around the time I got the lead in *The Chocolate Soldier*. At Hume-Fogg High, I reckon."

By the time the grown Randall joined Cal at Kenyon, he had left behind him Uncle Howl's "mash the button" and "pitch a fit" and "fuss you out." He had eradicated such mother-isms as "drawers" and "britches." And about all that was left—until Cal joshed him out of them—were his grand-mother's "Y'reckon?" and "I guarantee" and "I declare." *Each* of which Randall revived at whim for comic relief on the domes-tic front.

When Cal had served time as a CO and Randall was still serving time in the Air Force Celestial Navigation Tower, Ran-dall wrote him an exalting letter about the poems for his first book, *Lord Weary's Castle*, and he offered to "go over" the punc-tuation since "sometimes you punctuate as a matter of taste, but sometimes you punctuate the way the Irishman played the violin by main force." Cal seized the offer, knowing Randall was a Paganini of punctuation. Skilled in teaching freshman grammar, he knew in his sleep where any comma, semicolon, or period belonged. But, curiously, for the emerging romanti-cism in what Randall was calling his new "civilian poems," Punctuation 101 lacked something. These new digressive and self-focused poems hinted, rather than stated, their longings, and they needed a special punctuation for suspension, that is, a printer's symbol obliging the person reading the poem aloud to *pause in mid-breath* where a line breaks in mid-thought.

The aim here is to suspend the listener from passive listen-ing and rouse one to responsive listening. In that moment of suspension, one can be awakened to the poem's true underly-ing statement. Does one agree with it? Or disagree? Or leave it up in the air? Does one finish the thought with one's own thought? The suspension symbol is invaluable software for the romantic poet for whom subtlety is all. The dash-enclosed phrase cannot pull this off and the question mark attracts too much attention. Randall "discovered" that the wonder-worker, to both *engage* and *suspend* the listener, was the three-dot ellipsis and he taught it many tricks. He taught it to enhance three ti- tles so that "Come to the Stone . . ." became haunting; "When I Was Home Last Christmas . . ." enigmatic; and "What Is the Riddle . . ." mystic. He taught it to keep time with his scansion and in one case to make a little pun. In the poem, a Proust reader is lamenting that "a woman never is a man's type" and he goes on to say:

> *Possessed by that prehistoric unforgettable*
> *Other One, who never again is equaled*
> *By anyone, he searches for his ideal,*
> *The Good Whore who reminds him of his mother.*
> *The realities are too much one or the other,*
> *Too much like Mother or too bad . . . Too bad!*

"Woman"

What is that ellipsis up to? Is it "too bad" those realities are what they are? Or "too bad" the Good Whore is so bad?

Randall was enchanted with his new device and set it to aer- ating the longueurs in "The Night Before the Night Before Christmas" *thirty-one times*! A few maturing years later found him using it with restraint and cunning only twice in "Aging" and, of course, the ellipsis was indispensable with each nostalgic "*If only* . . ." and "*And yet* . . ." and "*Still* . . ." Eventually it became his rhetorical signature, as characteristic of Jarrell as a plucked string in Brahms. In his very late poem, "The Player Piano," the three dots perform their magic for him one last time.

In the fifties, serving at the Library of Congress, Randall was a willing circuit-reader for the Congressional Wives, the Naval Academy Faculty Wives, and a network of ladies' book clubs. I went along, too.

They were rapt listeners with eager faces, and Randall was at his gentlest with them, claiming them as the next best thing to students back home. Keenly sensitive to how hard it was for audiences to grasp all of a poem merely hearing it *once*, Randall chose his programs for the ladies with affectionate concern.

"I want them to *enjoy* the poems. Once over lightly is so unfair to them, poor lambs. And unfair to the poems, too."

He often introduced a poem with a few words to the wise, but assuming his mid-life listeners would *get* "Aging" intuitively, he simply explained the ellipsis. I can hear him now.

"Ah . . . ah, I think I ought to warn my listeners about this next poem. It's called 'Aging' . . . and . . . and, well . . . When I make my voice funny and look out at you, well . . . ah . . . fear not, all is intended." He would stop and smile at us.

"It just means that we've come to an ellipsis. You know, three dots?" Blank stares. In Washington, the Ellipse is a geometrical landmark. "Oh, it's like . . . it's like . . ." Teacher Randall persisted. "Ah, I know what it's like. Remember those scripts they'd give us for the high school play? Where they had in italics *dramatic pause*?" Then, relieved by their relaxing faces, he stopped "explaining"—as if he had—and said, "There'll be two dramatic pauses . . . two ellipses in the next poem."

Aging

I wake, but before I know it it is done,
The day, I sleep. And of days like these the years,
A life is made. I nod, consenting to my life.

. . . But who can live in these quick-passing hours?
I need to find again, to make a life,
A child's Sunday afternoon, the Pleasure Drive
Where everything went by but time; the Study Hour
Spent at a desk, with folded hands, in waiting.
In those I could make. Did I not make in them
Myself? The Grown One whose time shortens,
Breath quickens, heart beats faster, till at last
It catches, skips . . . Yet those hours that seemed, were endless
Were still not long enough to have remade
My childish heart: the heart that must have, always,
To make anything of anything, not time,
Not time but—
                    but, alas! eternity.

Uncannily, Randall was right. At his first ellipsis their chins went up, "*Consenting to my life?*" Yes. No. Whatever. They were into the poem. At the second pause when the Grown One's heart was skipping, they pulled up straight in their chairs, eyes bright with reckoning the Grown One's fate. Was it cardiac arrest? Or heartburn? Who among us in the tick of that ellipsis was not mindful of how far along *she* was in everyone's amble toward our end?

For complete alliterative and onomatopoetic musicality "The Lines" is a triumph. In it, Randall calls on all the techniques of the trade, using them with an artist's sleight-of-hand in just the right amount, in just the right place, at just the right moment to perfect the work and yet to make it seem effortless and easy. Again, as with "Aging," there is the automatic *self*-recognition with all the world who have ever stood at the end of the checkout line in the supermarket at rush hour. And yet, duded up in soldier kit the cliché crisps and freshens.

The Lines

After the centers' naked files, the basic line
Standing outside a building in the cold
Of the late or early darkness, waiting
For meals or mail or salvage, or to wait
To form a line to form a line to form a line;
After the things have learned that they are things,
Used up as things are, pieces of the plain
Flat object-language of a child or states;
After the lines, through trucks, through transports, to the lines
Where the things die as though they were not things—
But lie as numbers in the crosses' lines;
After the files that ebb into the rows
Of the white beds of the quiet wards, the lines
Where some are salvaged for their state, but some
Remanded, useless, to the centers' files;
After the naked things, told they are men,
Have lined once more for papers, pensions—suddenly
The lines break up, for good; and for a breath,
The longest of their lives, the men are free.

At first hearing, the rhythmic "line to form a line to form a line to form a line" is all that sinks in. It takes a second reading, or more, before we knowingly drift with the hypnotic euphony of the long "i's" in "file," "line," "side," "child," "die," "lie," "white," "remind," and "quiet."

So compact is the virtuosity of this poem that one can be dead and buried in it, symbolically, without ear-hearing or eye-noting the consummate wit of the punctuation. How those colons and commas grafted into the loose pentameter of eighteen musical lines *hinges them into one sentence.* One line.

I remember telling Randall this before we were married, when I was earnestly studying his poetry. We were in Laguna out on the Seal Rocks savoring the surf. The more I rambled on, the more Randall nodded encouragingly at me in a pleased-professor way. "It's so, child. But where got ye these truths?"

All at once we were scrambling up the rocks to escape a mounting wave and when we were settled high and dry again, he said, "You know, when I was in the army, I made a vow that when I got out, if I wanted to see a movie and there were three people *in a line* at the box office, I wasn't going to see that movie."

But of course, we did. When I knew him better, I learned that this was just one more sample of the reckless pronouncements Randall half-believed when he made them, only to look you in the eye six months later and say, "Really? I said that? I declare. Oh, I'm so silly."

A similar conversation took place years later when we were rusticating in our house in the Quaker woods near Greensboro. I was washing dishes and Randall was drying them and we had Berlioz on the classical music station. I remember Randall gazing contentedly at the new MacIntosh amplifiers music critic Bernard Haggin had imperiously advised. Then he said, "You know, pet cat, if I hadn't been a poet, I think I might have been a composer."

"Well," I went along with the game, "you'd have had to learn the piano, wouldn't you?"

"Oh, not necessarily," he said. "Berlioz didn't. He only played the guitar."

"Guitar! You'd have studied *guitar*?"

And Randall threw the dish towel over my head.

"Oh, you're so bad, bad, *bad*, von S."

Because I knew as well as he that Ear couldn't stand guitar. Or harpsichord. We owned no Laredo records and our one Wanda Landowska was for piano because, to Ear, solo harpsichord was "a hundred rubber bands around a cigar box full of bees."

Randall's composer fantasy was half-realized in that he poured a good measure of himself into Gottfried Rosenbaum in *Pictures*. It was no accident that Gottfried, his Viennese twelve-tone composer, played tennis, drove a European sports car, prized a cable-knit sweater like Randall's, and enthused over Arnold Schoenberg's *Verklarte Nacht* . . . "our" song.

Furthermore, had not Randall composed already his own poetic *Eine kleine nachtmusik*, "Che Faro Senza Euridice." "Farewell Symphony," "Rage for the Lost Penny," "Difficult: Resolution," "Scherzo," "Lullaby," "Variations," "Rhapsody," and "1938: Tales of the Vienna Woods"?

In a letter to his mother from his lost world of Los Angeles, thirteen-year-old Randall had written, "For the last two nights we've been getting grand opera on the radio. If you buy a ticket it costs *five dollars* but we get it free. It sure is nice. It's lots better than jazz. It's hot stuff."

Ear was permanently repelled by the nasal quality of the jazz saxophone "like a kazoo with a head cold," Randall said, and equally repugnant was its jazz companion: the jazz snare drum. To Ear, such rat-a-tat was so offensive that one evening at the Staatsoper in Vienna the Jarrells ducked out on Donizetti's *Daughter of the Regiment* with Simianato.

In another letter from California to his mother, Randall wrote, "Sunday we saw four great big buzzards right in the road eating a dead chicken. They just stalked to the side of the road when we passed and stalked back again. They sure are mean, ugly-looking birds. They just sail around in the sky looking for something to eat. They seem to say 'we'll get you someday, get you, get you, we'll get you yet, get you yet, get you yet.' Just like choruses of songs that run together."

According to his mother, Randall's sensitivity to sound began in his infancy. She said, "Ran's first words weren't *Ma-ma* or *Da-da,* but *Bow-Wow.* Talkin' to the puppy in the other side of the duplex. Lil' Spitz dog." Years later, Ear still heard talking animals. Our yard cardinals said, "Birdie-birdie-birdie" and our cats did not meow above their dishes, but said, "Now! Now!" On the other hand, when I talked back to Kitten or Elfie in that whine-snarl they argued in, and which my daughters thought I did rather well, Ear flinched and Randall beseeched, "*Please* don't make your voice like that, best beloved."

Ear enjoyed my imitation of the mynah bird we met in a pet shop in Santa Barbara. He used to clack at us in prosthetic monotone, "I-am-your-friend. I-am-your-friend." I did my mynah bird whether asked or unasked, and I happily complied with requests for my TV commercial of a schoolchild's enunciative treble hype for "Peter-Pan Peanut-Butter. It's the *Peee-*nuttiest!"

Trendy slang like "fridge" and "cukes" grated on Randall's nerves. Once (horrors!) I said "Jaggy-boo." *Once.* But just let my mother speak of "scraggly neck" or "skowy skirt" or "ta-mah-toes" and Randall was all suppressed smiles. Soul-talk was music to his ears. He treasured—and brought back to me—our black helpers' best phrases as he drove them home, like, "We uns mos' dere" and "M'daddy cain't half-hear" and "Don' pay it no neva mine" and "It's de fuss nighta de roun' moon." When I reported that the yard man had asked me, "Mista Jurl? Him a book-ratter?" Randall nearly swooned with bliss, "That's what I want on my tombstone," he cried. "I guarantee!"

Tuning in to the classic music station on the car radio, Ran-

dall could identify, at the first notes, and keep Poulenc, Boc-
cherini, Elgar. Or else flip the dial saying, "Away, away base
Menotti," or "Dopey old Hindemith." Virgil Thomson's
Louisiana Story Lyric Suite also got the flip, along with Randall's
epithet "The Battlemarch of the Alligators!" Sometimes he
might pause and say to the radio, "Hmm . . . who . . . who, pray,
is this? German . . . Late romantic . . . Mahler? No . . . *Early*
Mahler? Some so-so composer. Oh, I know . . . I know. I know.
It's Bruckner. Probably the Ninth."

"*How* do you know?" I'd ask.

"A breeze, little Peterkin. A nothing." He would pat my
knee happily. "Whenever it sounds like inferior Mahler, it's
Bruckner. And it's usually his Ninth Symphony because that is
all they know about Bruckner in this country. See how smart
your Randall is?"

Afternoons in our woods, he writing in his hammock, I
reading in mine, were idyllic before the developers tuned up
their chain saws. Then, at the first ugly snort, we leaped from
our hammocks, grabbed our binoculars, and drove out in the
cornfields to look for cowbirds.

At the sports car races in Palm Springs, when a few motor-
cycle laps were tolerated between the heats among Europe's
finest, the ear-splitting rip of their gearing was a dreaded torment to Randall. Impossible to speak above their deafening ac-
companying roar, Randall could only clench his jaw, clamp his
hands over his ears, and *suffer in silence*. Just the opposite of this
was the calm, even purr of the dual exhausts on our Jaguar
XK–120. Setting out for class, I at the wheel and Randall standing by to close the garage doors, I could glimpse his tilting,
thoughtful face transfixed—as if hearing *Elektra*—by the elegant hum of the costly valves and cylinders. Sometimes he
climbed in beside me murmuring "Mama. *Mama.*" Sometimes

he'd take my place at the wheel and tell me just where to stand in back so I could listen, too.

Apart from the tormenting *sounds* Ear heard from harpsichords, motorcycles, and my caterwauling, there were, for Randall, words of intolerable offense ("ass," "bullshit," "crap," "piss," "fuck," among others) that were popularized by Ginsberg and followers and were infesting the taste and language of Randall's students and colleagues.

"What a way to talk, or write," he fumed at his poetry and writing classes as a shaming scowl creased his face.

"I mean compared with Peter Taylor . . . with F. Scott Fitzgerald. With Frost. Or Eliot. Or Gerard Manley Hopkins." Then, sorry to scold his lambs for long, Randall would sign off, all wounded appeal, saying, "I mean . . . well, *golly*."

As a teacher he longed for them—for their own good for the rest of their lives—to prefer genuine literary style, to scorn the novelty coarseness of Beat chic. As the private and discriminating person Randall was he enjoyed identifying himself with Samuel Johnson's phrase, "I consider myself an ornament of society," and he occasionally quoted it—half in earnest. At heart, those close to Randall felt he yearned for that from all of us. Certainly no member of our household spoke in obscenities. The worst epithets Randall used for certain people or certain situations were "stupid," "dopey," "idiotic," "half-witted," "pigheaded," and so on.

Peter's language in his writing and in our company was entirely acceptable for Randall and little old ladies. Cal, always sensitive to falling out of Randall's favor, endearingly refrained from all Ginsbergian inclinations in Randall's presence, even to speaking of "bull" without the suffix.

Randall sang not, neither did he whistle. Ear detested whistling. All he played was his stereo: and favorite instrument? The piano. He had Lipatti, Anda, Cziffra, Skoda, Bolet, and everything obtainable—in the fifties and early sixties—of Richter and a newcomer, Brendel. Browsing among our pianists for whom to play for the mood we were in, Randall sometimes sighed and said, more than once, "You'd have thought *somebody* would have gotten me piano lessons."

Somebody of course was Uncle Howl who bankrolled the basics for the single-parent household but thought in terms of jobs at his candy factory, rather than piano lessons and a piano. Joking about this with Randall later, Uncle Howl laughed merrily, saying, "Oh, we might coulda, I reckon. But we din know back then Ran, you gonna getcher name in the paper."

In 1948, Randall saw Europe for the first time; heard Mahler, and Richard Strauss operas for the first time; and wrote James Laughlin, "Europe had about as much effect on me as the Coliseum had on Daisy Miller." In fact, Randall's discovery of Mahler and Strauss was what opened Ear to twentieth-century music.

In particular, the Austrian summer prompted ten poems about Europe, which with eighteen other postwar poems made his civilian volume called *The Seven-League Crutches*. Furthermore, these European ten were written to the musical accompaniment of Mahler and Richard Strauss—instead of his old favorites Haydn, Mozart, Bach, Brahms, and Beethoven—and in their way they established for all his poems to come a new looseness in form, a new intimacy of theme, and a new cultural richness in tone.

In a letter to his Viennese friend Elisabeth Eisler, he noted Hannah's praise of "my new way of writing" and added, "I've never written such poems before." Another letter to her said, "remember I once said, scornfully, that I wasn't going to the symphony to hear Mahler? Well, I now feel terribly foolish to have said so, and he's become one of my favorite composers. I've got some of his symphonies from the record library at the college . . . and by now I know the First, Second, Fourth, Fifth, and *Das Lied von der Erde* well. He seems wonderful to me, and his neglect in this country seems particularly strange. I'm sure that in the end he will be thought a great composer by everybody."

Eventually we bought all the symphonies Bruno Walter conducted and all the Schwarzkopf songs. The "dew-hushed drums" in his poem "Nestus Gurley" referred to Mahler's great, velvet-toned kettles, Randall said. And the lines:

> *When, somewhere under the east,*
> *The great march begins with birds and silence;*
> *When, in the day's first triumph, dawn*
> *Rides over the houses . . .*
> *The Complete Poems*

are what, between ourselves, we called "the Mahler part."

Listeners and interviewers were always puzzled at Randall's lack of Southern drawl or twang. Sooner or later, they would question him:

"Weren't your family Southerners?"

"Through and through."

"And weren't you born in Tennessee?"

"That's right." Randall was amiable but had to tease, "Really though I learned to talk in California."

At home by ourselves with Elfie, I reminded him he had once told me he "made up his mind" not to talk like that and it was simply an act of will. Randall looked at me queerly and said,

"The things you remember . . ." Before I could answer back, he added in self-defense, "Well, at bottom . . . how could I possibly teach Eliot and sound like Uncle Howl? It had to go. All of it. Old Cal was right."

Old Cal, sure. But really it was Ear.

VII

Proust says that intelligence and sensibility are rarely accompanied by will; and this (if I may immodestly give myself intelligence and sensibility) is sadly true of me. It's very hard for me to force myself to do anything unpleasant or dreary. (If I'm forced to, as I was in the army, that's different.) I guess I can sum up my bad points by saying I can't, even if I try, be dutiful and make my life careful and methodical and unselfish and self-sacrificing. Even if I tried, I wouldn't succeed.

O, and I forgot to say, I am childish in many ways, but this is as much good as bad.

RANDALL JARRELL

From a letter to Elisabeth Eisler

On Receiving the National Book Award: Manhattan

ELLIOT ERWITT

The Children's Quartet

RANDALL SPENT MOST OF HIS GROWN-UP LIFE writing for grown-ups. Out in the world, Randall was mainly known as the author of the most anthologized poem of World War II, "The Death of the Ball Turret Gunner," and known in our town as the bearded professor who played tennis in the city tournaments and drove a Mercedes convertible.

In February 1962, Randall was hospitalized for hepatitis and confined to bed too weak to read his mail. A letter from a Michael di Capua written on Macmillan stationery was among the get-well cards I showed him, and it was the turning point in Randall's recovery and a final turning point in his literary career.

As I read that letter aloud, Randall's eyes lighted up for the first time in weeks. To his joy, this di Capua knew Randall's poetry and spotted his affinities with Goethe, Rilke, and the Grimms. Di Capua asked him to translate five Grimms' tales for an unusual series of children's classics. This series was the first brainchild of di Capua—a young, newly hired member of Macmillan's children's literature department. He was matching such writers as Jean Stafford with *The Arabian Nights*, Isak Dinesen with Hans Christian Andersen, and John Updike with Oscar Wilde. Randall sat up in his bed and said, "Write him back for me, will you, pretty secretary? Tell him I'm sick now, but I'll start on them as soon as I can. In fact, how about stopping by the library and bringing me a collected Grimms'? In German, *naturlich*." Randall smiled as he leaned back to relax, "What's this guy's name?" he asked. "Michael di Capua," I said and we both shrugged.

"Well," Randall added, "he's a mighty cultivated reader." For years he had been hoping for an editor who was well read, who knew his work, and who would have plans—enthusiastic plans—for it, an editor who would answer his letters by return mail and *not* be a writer himself. In the months ahead Randall found that di Capua was this person.

Randall's first contribution to the Macmillan series was titled *The Golden Bird* and consisted of five Grimms' stories and a short introduction for them. A second commission followed and was titled *The Rabbit Catcher*, for which Randall translated three stories by Ludwig Bechstein: "The Rabbit Catcher," "The Brave Flute Player," and "The Man and Wife in the Vinegar Jug."

Each completed translation brought an immediate long-distance call from a gentle voice in New York full of appreciation and interested questions—a voice that belonged to someone with all the time in the world, it seemed, to listen. Soon, gift books began to arrive on Gogol, Hardy, Richard Strauss, and whatever else the attentive di Capua noted Randall was interested in. Letters came every week, written in a clear and readable longhand and reporting progress on the books, plans for an enlarged edition of Randall's translations, and entertaining bits of news along the Rialto. Later on, there were requests from di Capua for a weekend date when we would come to New York and the promise he would get us—God knows how—opera tickets. When we first met our telephone friend in his cubbyhole at Macmillan's, he was shorter and younger than we expected and *seemed* less commanding than Randall's other editors. Little did we know that behind those inexpensive spectacles and under that thinning head of hair was a Madison Avenue *wunderkind* who would be, at thirty-five, a senior children's editor revered and feared the length and breadth of the trade. On our flight home, Randall and I marveled at di Capua's attention span when Jarrell talked, the way he treated Randall as an end in himself—a person, as well as a writer with possibilities—and at the feeling he gave us of being on our side. We both laughed in mild astonishment recalling di Capua's mini-tantrum with the maitre d' and his tough treatment of taxi drivers and beggars. We liked him. Many opera weekends later, on a brisk, sunny day in New York, we were walking—three friends abreast—down Fifth Avenue to di Capua's newest restaurant discovery. Randall was saying ruefully what a long time it had been since he had written any poems and at that di Capua made his move. "What about writing for children, Randall?" he asked so smoothly. "Have you ever thought of that?" And Randall Jarrell, an author of juveniles, was invented.

In Greensboro, after his classes were over for the day, Randall and I took ourselves to the children's section of the downtown library. In the little chairs at the low tables, with a stack of books before us, we began leafing through them "to get the feel of the genre," as Randall quipped.

"So, that spring I often lay in our hammock outdoors," Randall told Aaron Kramer in a radio interview in New York, "and wrote every day on a little story about a rabbit, and I'd read what I wrote to my wife who was gardening there. I enjoyed it. It wasn't a 'real' book . . . But it was fun."

Although this was not poetry, it was fun in that it was something of his own in place of endlessly translating *Faust*. It did not feel like a "real" book because in his innocence of the genre Randall had underrated it and cooked up *The Gingerbread Rabbit* the way a master chef cooks up something playful for his child on a day off. Those familiar with Randall's writing for grown-ups will notice how he stuffed and spiced *The Gingerbread Rabbit* with familiar and not-so-bland Jarrell ingredients but sugared them over for the palate of a child of five.

Gone with the wind is Randall's old nemesis, the black-breasted witch-mother of his poems "The House in the Wood" and "A Quilt Pattern," who fattens her boy to roast in the oven. In *The Gingerbread Rabbit* the witch-mother is replaced by an angel-food-cake-mother who bakes savory surprises for Little Mary after school.

Likewise the rabbits in these pages are cozy and adorable—as nowhere else in Jarrell—and treated as fluffy comfort symbols, similar to his own pet rabbit given him by his grandparents the year he stayed with them in Hollywood. Nowhere in this book is the usual caged, helpless, victim-symbol rabbit that hops feebly in "Stalag Luft," "A Bird of Night," and "A Street Off Sunset"; nor, as in "A Child of Courts," the rabbit that dies "for a use"—the way Randall's own childhood rabbit did after the boy returned to his mother in Nashville and his down-to-earth grandparents made a meal of the pet.

At first the babyish, guileless, doughy gingerbread rabbit seems destined for the same cruel fate but, after giving him a fright or two, the author sets him on the road marked Happy Ending. There, the orphan-hero is saved from the jaws of a fox by a fatherly "big, brown rabbit" who, with "Darling," his wife, wants to adopt him and have him sleep in "a little rush bed" and feed on "tiny, golden carrots." In the end, their offer brings tears to the gingerbread rabbit's little raisin eyes, and he exclaims, as Randall did, vainly, to his grandparents, "I'd like to live with you always. Always!"

This fantasy of stepping into someone else's already running household is often typified by Randall's adoption themes, as in "A Sick Child," in which the child wants some beings from another planet to come to him. It is fully developed in "Windows," where a man on a snowy sidewalk looks through the living room window of a lighted house and longs to be there with the woman who is darning while her husband nods in his chair. In his loneliness the man thinks of entering through one of the many windows.

> *Some morning they will come downstairs and find me.*
> *They will start to speak, and then smile speechlessly,*
> *Shifting the plates, set another place*
> *At a table shining by a silent fire.*
> *When I have eaten they will say, "You will have not slept."*
> *And from the sofa, mounded in my quilt,*
> *My face on their pillow, that is always cool,*
> *I will look up speechlessly into a—*
> *It blurs, and there is drawn across my face*
> *As my eyes close, a hand's slow fire-warmed flesh.*

> "Windows,"
> The Complete Poems

Later, in his novel *Pictures*, the innocent and guileless girl student at Benton College is, in the end, *adopted* by the childless Rosenbaums, a professor and his wife. And there the author cannot resist intervening in the story, asking them, "Aren't you going to adopt me, too?" Again, in the end of Randall's children's book *The Animal Family*, the boy washed ashore on the island is *adopted* by the hunter and the mermaid.

The Gingerbread Rabbit—a book where the animals talk—is the natural outcome of a writer whose favorite juvenile reading, along with Hawthorne's *Tanglewood Tales*, included *Aesop's Fables*, *The Jungle Books*, and *Uncle Remus*. That the dialogue is the outstanding feature of the book is another natural outcome from a writer long used to treating dialogue as an art form in his prose and poetry. Particularly exciting to a child are the seductive lies of the fox and the menacing lines of the utensils as they make sport of the doughy rabbit spread out on the kitchen table while the oven heats. They seem sufficiently entertaining in themselves, but are even more so if the parental reader can detect their parallels in the sinister "house" that talks in "A Quilt Pattern" and the wily Mephistopheles in *Faust*.

There is something else about *The Gingerbread Rabbit* that is not likely to be known by those outside the family. Randall endowed Little Mary's mother in the story with two distinctive characteristics of his own: his delight in giving and in surprising. A week rarely passed that he did not come up to one of us with anything (from the latest Schwarzkopf recording to a red star-shaped gum leaf) and happily give us the mock command:

> *Open your hand,*
> *And close your eyes—*
> *And I'll give you something*
> *To make you wise,*
> *Surprise,*
> Surprise!

Concerned primarily with the writing of this first children's book, Randall rather assumed that the writer found some illustrations he liked, told the editor about it, and the illustrator would come running. After leafing through more stacks of books on the low tables and in the little chairs, Randall found that Garth Williams's pen-and-ink drawings of the furry animals in Margery Sharp's *The Rescuers* were just what he wanted. Turning the matter over to di Capua Randall had no idea of the difficult task he was setting for a very junior assistant editor at a very early stage in his career: to induce the established and much-in-demand illustrator of, among others, E. B. White's *Charlotte's Web* to take on a first book by an unknown children's writer. Di Capua kept to himself any qualms he had and concentrated his powers of persuasion on winning Williams over. That done, di Capua became an accomplished go-between for the two older artists—one in North Carolina, one in Mexico—who never had any direct personal contact with each other.

So it was that in July 1962 Randall sent to di Capua—to relay to Williams—a letter of guidance, saying:

> I'll write out a few things about the characters in *The Gingerbread Rabbit* and the most likely things to illustrate. The rabbit himself ought to be very sincere and naive and ingenuous, so that his whole body and face express what he feels. The big rabbit ought to be handsome, secure and competent-looking; the mother rabbit should be delicate and demure and beautiful. The fox should be very smooth and flashy, like Valentino playing W. C. Fields. The little girl's mother should be young (28 or so) and beautiful and kind, just the mother a little girl would want; the little girl should be something any little girl can immediately identify with.

Eight months later, in March 1963, Randall wrote di Capua: "I am delighted with the drawings: the gingerbread rabbit's *very* cute and touching. The fox is wonderful, and the old rabbit in the colored sketch makes me want to be *adopted* by him. . . . I believe Williams is getting quite inspired and will make a charming book."

After some astute inquiries from di Capua about what he would write next for children, Randall explained that he was busy preparing a formal lecture for the National Poetry Festival in Washington to be called "Fifty Years of American Poetry" and writing anthology introductions for Kipling and a volume of Russian literature. "But I'm sure I can do some more things for children during the winter and spring," he wrote. "Doing them is a *great* pleasure and, so far, I don't have any of the trouble with them I have with grown-up things—may Heaven keep me in that state always!"

II

By November, as Randall had hoped, he was able to start another book. He later described it to Aaron Kramer as:

> "a book half for children, half for grown-ups, called *The Bat-Poet*, that felt just like a regular book to me. And, you know how it is, you work on it all the time. You stay awake thinking about it at night. You wake up in the middle of the night . . . and, I did it just like a grown-up book. By good luck, we have some bats on our porch. I think they like the insects that come to the porch light. And anyway, there really was one bat that was a different color from the rest. He was a kind of café-au-lait brown and I made him the bat-poet. And so I imagined a bat who could not *write* poems, but who would make them up; and so I had to make up poems for him. And a couple of the poems ["The Mocking-bird" and "The Chipmunk's Day"] were pretty much like grown-up poems—anyway, *The New Yorker* printed them. I didn't tell them they were children's poems."

The story the grown-ups hear when they read *The Bat-Poet* to their children is the allegory of the artist who, in Goethe's words, has "two hearts in one breast." They find, through the bat, not only his poet's heart, which dares to turn from the conventional world to a self-invented one, but also his creature's heart, which is lonely for his fellows to share in his world

101

of art and imagination. But are his fellow bats ready to enter into, or even hear about, a foreign country with abrasive sounds and highly evolved monsters, and charged with an incomprehensible dimension called Color? No. They are afraid and, though they are polite, they flee from the bat as a dubious and eccentric character. While the bat tries to resign himself to their indifference and to enjoy his reception among strangers, the time comes when he, too, must hibernate as the other bats have done already. The happy ending for him is that he was able to write a poem (not about the daylight world, but about bats) and that his new friend (the chipmunk) says the bats are *sure* to like it "the way I like the one about me."

The story the children usually hear is a simpler one, a story more explicitly concerned with a bat who writes poetry. His friend, a practical-minded chipmunk who likes the bat, tags these poems "portraits in verse" as he markets them in the natural world.

On the other hand, this book "half for children and half for grown-ups" is really for artists: that is, the book tells the rest of the world what it is like for a tiny percentage to want to be part of the whole—not isolated; to communicate with the great world through literature, painting, music, and the other arts; and finally, to make the world understand that art for the artist is more than turning on the spigot.

A good example of this occurs when the bat is unable to compose a poem about the cardinal. Since the bat had no difficulty writing about the mockingbird, the chipmunk, and the owl, it might be assumed he could write about the cardinal—especially since he admired the cardinal and was interested in him. It was puzzling to the chipmunk that the bat could not; for that matter, it puzzled the bat; but no matter how much he wanted to do the cardinal's portrait, his vision failed him. This minor episode in the story came from Randall's attempt to write a book on Hart Crane for Holt and Company. Although Randall admired Crane, was extremely interested in him, and had accepted a $2,000 advance to write about him, he could not. After struggling more than a year with notes on Crane, and after spending the advance, Randall had to admit, as the bat did, "I would if I could, but I can't. I don't know why I can't, but I can't." And in Randall's case he had to struggle even further to pay back the advance.

Of the four portraits the bat accomplished, "The Mockingbird" was Randall's favorite and he placed it second in his table of contents for his later book *The Lost World*. Under its Wordsworthian surface and its Yeatsian conclusion, "The Mockingbird" is a caricature drawn from Randall's knowledge of Lowell's and Frost's, *and his own,* self-obsession, acute sensitivity, and fierce territoriality. Through this poem he is pleading for the world to see that these are part of the cost of being "a real artist"—but a negligible part and a pardonable one when compared with the mockingbird in his best moments, when the artist achieves his unsurpassable imitations of life that make us pause, look up from what we are doing, and say, "He's right. That's the way it really is."

The bat-poet himself has just this effect on the chipmunk in his portrait of the owl, "The Bird of Night." Randall based the poem on a barred owl we knew, one that hooted and hunted in our woods. Although the owl is first presented visually, it is plain, suddenly, one is hearing the owl, too. In a subtle onomatopoeia Randall creates out of repetitive "l's" and "w's" and legato vowels the stir of air when a sizable feathered object glides through it downward, and more downward to seize its prey. This imitation of life was so real to the chipmunk that he shivered at the end and said, "It's terrible, just terrible! . . . I'm going to bed earlier. Sometimes when there're lots of nuts I stay out till it's pretty dark; but believe me, I'm never going to again."

The chipmunk sat twice for his portrait. Once he is depicted in verse—also onomatopoeic—by means of short, staccato words and lines and the musicality of "o" and "e" sounds punctuated with sharp "t's." This portrays him as a neat, fleet body darting through his days, content at sunset to have fed himself and escaped the talons of the owl. His prose portrait was a character study revealing him as the bat's devoted friend, loyal supporter, and able entrepreneur. This I recognized—and Randall admitted laughingly—was a sort of tribute to di Capua and to me.

The facts for the family portrait "Bats" were gleaned from Randall's reading about their mammalian birthing and suckling behavior, and the portrait was modified by his summerlong observation of the vulnerable little colony bunched upside-down by our porch light. By now quite fond of the bats Randall approached the poem in a lyrical way, and somewhat transformed their homely truths into such lines as these:

A bat is born
Naked and blind and pale.
His mother makes a pocket of her tail
And catches him. He clings to her long fur
By his thumbs and toes and teeth.
And then the mother dances through the night
Doubling and looping, soaring, somersaulting—
Her baby hangs on underneath.
All night, in happiness, she hunts and flies. . . .

Her baby drinks the milk she makes him
In moonlight or starlight, in mid-air.
Their single shadow, printed on the moon
Or fluttering across the stars,
Whirls on all night . . .

"Bats,"
The Complete Poems

Randall simply assumed Garth Williams would do the illustrations for *The Bat-Poet*. So absorbed in the new poems he was
able to write, Randall had not given the illustrator a second
thought: but di Capua had. Di Capua had his eye on young
Maurice Sendak, who was a versatile and original artist with a
growing reputation, who had his own best-seller at Harper's
(*The Nutshell Library*), and who was, furthermore, a Grimms' en-
thusiast. Di Capua proposed Sendak for *The Bat-Poet* and in the
future for an expanded volume of Grimms' tales to be translated by Randall.

At first this was a jolt. Sendak was not a household name in
North Carolina and Williams was—in our household at least, especially due to his association with *The New Yorker* writer E. B.
White. This time di Capua focused his powers of persuasion on
Randall, and after another stint in the little chairs at the low tables Randall was able to write him, "[On seeing] nine or ten
Sendak books and five or six more of Garth Williams' books, I feel
as you do about Sendak. He's better at lyric or quiet or thoughtful or imaginative effects. . . . I liked many of his animals—his
bear family, for instance—very much. He would probably be
better for the tone of *The Bat-Poet*." Thus, the decision was made.

In a "Dear Mr. Sendak" letter of February 1963 Randall said:
"I've almost no suggestions and want you to do it just as it
comes to you. The animals like the mockingbird and chip-

munk are very much the same as the real ones, so color photographs or watching the real animals might help you. That
particular sort of little brown bat, on the other hand, has too
much a devil's face to use; other sorts of bat, just bats for instance, have faces more like squirrels' or mice's and you could
invent a face for him more like theirs." Randall closed with,
"We had an awfully good time at your house after the Opera; I
hope you can come visit us sometime. In the spring the bats,
the chipmunk, the mockingbird and the cardinal are all right
here."

Though Sendak was not able to visit the bat-poet's home, he
was able—by some inexplicable clairvoyance—to duplicate,
almost, the exact corner and woodwork and light fixture on
our porch where the bats clustered. Further on in the book,
again coincidentally, he created an uncanny likeness in the
wooded scene with the old-fashioned bench to our own
mossy, wild, side garden.

When Randall saw the finished book he wrote di Capua: "I
(Mary, too) was really *enchanted* with the look of *The Bat-Poet*. It's
exactly right, I think; it even has a sort of serious, elevated look
as if it were a classic, as much for grown-ups as for children. I
think Maurice did it wonderfully. The printing and the paper
are perfect. I'm crazy about the small bat drawings at the side
[of the pages] the ones he [added] at the last; some of them are
the most accomplished drawings of bats I've ever seen." Randall was also pleased with his own part of *The Bat-Poet* and said
in the Kramer interview, "I've been awfully happy about what
readers have said to me and what I've read in the reviews. You
know sometimes you feel you have good *luck* with a book.
Things *come* to you. And I feel that way about it, . . . that *The Bat-
Poet*, for what it is, is done right."

III

Having written the four poems for *The Bat-Poet* Randall wrote twenty more during the next two years, which made possible his last book of poetry, *The Lost World*. At the same time he was writing not one, but two children's stories: *The Animal Family*, which di Capua took with him when he left Macmillan for Pantheon, and *Fly by Night*, which di Capua took with him to Farrar, Straus and Giroux.

The very first intimation of *The Animal Family* as a book dates back to 1951 when Randall was teaching at Princeton, getting divorced, and trying to keep house alone. In November of that year he wrote me in Laguna, California: "I think I'm going to write a poem rather like 'Deutsch Durch Freud' called 'The Poet-Cook'; at least, I've been thinking about it. I'll have a mermaid in it that comes to live with the Hero. I've already thought of some details like little pools of water where she stands, [and] the flop of her tail as she comes up the stairs. In some ways she's going to be like another mermaid I know—I bet nobody's ever written a poem about a poet-cook with a mermaid. Ah, originality!" Actually Randall's novel took immediate precedence over this poem, and there were other attractive options. Twelve years went by before he thought of the mermaid and the Hero again, but not as a poem. This time the little pools of water and the flop of the mermaid's tail went into a story and the Hero was not a poet-cook, but a hunter, and they all made a prose book called *The Animal Family*, dedicated to Elfie.

This joyous tale is about a Garden of Eden without the Expulsion, and it took place in Randall's mind partly along a wild shoreline "where the forest runs down to the sea" which we had seen in Oregon one summer and in a cove near Seal Rocks, in Laguna, where we swam with Catalina Island on the horizon. In the story the hunter's house was built by his own hands, but his deerskin rugs *really* came out of our own house and had been bought in Salzburg. Likewise, "the big brass horn he had found in a wreck" was one we found at Gucci's in Florence and hung over our mantel. The ship's figurehead that the mermaid brought up from the sea was based on a charming lead figure Randall bought us for us from an antiques dealer in Amsterdam. Its description of her for the book is "a woman with bare breast and fair hair, who clasped her hands behind her head; she wore a necklace of tiny blue flowers, and had a garland of big flowers around her thighs. But her legs and feet weren't a woman's at all, but the furry, delicate, sharp-hooved legs of a deer or goat—and they were crossed at the ankles."

The characters in *The Animal Family* were composites, of course, modeled on fact and fantasy. The hunter was created out of the part of Randall that had a beard and a fur hat made of coon tails, the part that still wanted to play with bows and arrows, and the part that fondly identified with Orion, the hunter constellation. In a letter in 1943 to his first wife, written from an airfield in Arizona, Randall described the night sky outside his tent as "extremely beautiful, and old Orion, my mascot (I'm convinced God made him for me) looks as H. G. Wells-ish as ever." Later "Orion's" starry belt and sword appeared in two poems, the novel, and a piece of criticism.

The mermaid of *The Animal Family* was Randall's ideal of a beautiful consort who, for love of him, would forsake her family and friends, learn a new and intellectual language, be dear and funny always, put him first, and never turn into a Wife or Mother.

Their silver-furred and silver-eyed lynx was, from the outside, a replica of the Canadian lynx we knew in the Washington Zoo. Underneath, however, the lynx was Randall's Kitten, and he was given Kitten's attributes—"delicate," "deft," "quick," and "clever." When the lynx kneads his paws on the chamois shirts or pounces on a ball or stretches out flat on his outstretched paws watching, thinking, and dozing, Randall is commemorating Kitten. The bear was brought in for comic relief and because Randall had a wish-fantasy about owning a bear. The boy joined them to complete the archetypal family, to satisfy the hunter's troubled longing for continuity, and to aesthetically round the circle that began when the hunter as a little boy was shipwrecked on the island.

The Animal Family was so directly inspired by the rocky coastline, the blue and white surf, and the infinitude of sky on the westward edge of our continent, that Randall thought nothing but nature photography would suit the story. After reading it, Sendak said it was an impossible one for him to illustrate, that "the images are so graphically created in the writing that Randall does not need me." As Sendak enthusiasts well know, his first principle is that illustrations must amplify the text and not duplicate it, and the vivid writing in this passage of Randall's is typical of what Sendak was up against:

> In spring the meadow that ran down from the cliff to the beach was all foam-white and sea-blue with flowers; the hunter looked at it and it was beautiful . . . And when at evening, past the dark blue shape of a far-off island, the sun

sank under the edge of the sea like a red world vanishing, the hunter saw it all, but there was no one to tell what he had seen.

Randall and I began searching through Ansel Adams and Edward Weston albums of Western photography, and Randall wrote di Capua to see what he could find in the New York Public Library. What he wanted, Randall said, was:

1) The coast seen from out at sea. 2) The normal view of the beach, with the surf, the meadow, and the mountains. 3) The view from above that the hunter and the mermaid see in the last scene, just before they go into the cabin.

Sendak was appealed to when nothing usable came out of our search, but he still maintained the impossibility of his illustrating such a book as *The Animal Family*. Randall knew what Sendak meant and was dismayed; di Capua knew, too, but for Randall's sake determinedly (but diplomatically) persuaded Sendak to "decorate" the book instead of "illustrating" it.

In *Fly by Night*, the poem "The Owl's Bedtime Story" is about a lonely owlet waiting in a hollow tree and thinking to himself, "Come home! Come home!" when his mother is off hunting as Randall's mother used to be off working. After a while, the owlet gets the courage to try his wings by day to rescue *(adopt)* an orphaned owlet whose cries he has heard. In the end, the "good" mother brings the nestlings things to eat and "when she opened her wings, they nestled to her breast."

David, the boy in *Fly by Night,* could fly by night and was lonely, too, in his hollow room down the hall, cut off from his parents. In his dream of flying—"really, he is floating"—David sees his mother and associates her with pancakes for breakfast, sees his father and associates him with Oedipal rivalry, and then floats out of the house to observe some mice below him and then some rabbits, sheep, and ponies. In classic dream tradition, David is given a dreamer's omniscient overview and the familiar paralysis, being unable to speak or act. Contrarily Randall gave him no stress or anxiety—only a kind of puzzled wonder. Never stating this—or anything else—with intensity, Randall steers the floating boy back to the security of his blanket where David "starts to fall asleep." The book ends in the daylight of the sunny kitchen where David's mother kisses him with a loving look "like . . . like?" Like the look of the mother owl, but that instantly dissolves into his own mother's look and she says to him, "In two shakes of a lamb's tail I'll have some pancakes ready for you."

Randall wrote Sendak concerning the pictures for *Fly by Night* "will be so easy for you to illustrate that I've laughed over the thought again and again. . . . Paragraph by paragraph it divides into pictures, and pictures thoroughly in your own style." While Randall had reached back into his own boyhood for the story, Sendak reached back into nineteenth-century England for the pictures and there, in a pastoral Hardy-ish countryside, Sendak rendered the late twilight and deep night that the nude, sleeping boy floated through.

In all his books for children Randall minded the golden rule that insists on a happy ending, and he did this in his dreambook in double measure: once at the conclusion of "The Owl's Bedtime Story" and then again in the finale of *Fly by Night* itself. With only one main character and hardly any plot, Randall made the technical challenge of this book the portrayal of a dream-state as much as that of a dream's contents.

Actually, Randall often evoked dreams as a fifth dimension and for years had used the dream device in poems as a way of entering the unconscious of his soldiers, women, and children; and more than once Randall's dead speak through dreams.

Scholars consistently list "dreams" as one of his six literary subjects and, in fact, four of Randall's poems have the word "dream" in the title and more than thirty of them are—or are about—dreams. Minute but typical is the allusion in "The Death of the Ball Turret Gunner":

> From my mother's sleep I fell into the State,
> And I hunched in its belly till my wet fur froze.
> Six miles from earth, loosed from its dream of life,
> I woke to black flak and the nightmare fighters.
> When I died they washed me out of the turret with a hose.

The Complete Poems

Dreams frequently came to his aid for a lyric or satiric metaphor in his criticism or as a mine for hidden truths about a character. Considering this, it is hardly surprising that Randall found the idea of creating a book-length dream irresistible and that, having got David's dream under way, made up one for David's mother, father, dog, and a whole flock of sheep—where Randall joked, "All of them except one are dreaming they're eating; that one is dreaming he's asleep."

Randall and Sendak knew, as you and I know, that "flying" has a sexual connotation and that such words as "fur," "rings," "panting," and "mice" have, for the Freudian initiate, an erotic charge; but the triumph of this book is that David knows none of this at all. Through Randall's delicate rhetoric and his sensitive tone, *Fly by Night* is a tranquil improvisation on the dream technique of tentative suggestion and elaborate disguise.

In a letter written in September 1965 Randall said:

Dear Maurice:

We loved everything you said in your letter and were so touched and pleased by the gift—of all the small drawings of the bat at the side of the page, that was the one I like best. Thank you!

I'm so glad that things have got better for you and that you're as you say, feeling yourself again. It's been so wonderful to be feeling myself again and to be living at home with Mary instead of nowhere in a hospital. I know how you feel about the bunch of gray hairs—I seem to have collected some myself. But God must be on the side of gray hairs, he makes so many of them.

I feel so lucky and grateful to have had your pictures for both The Animal Family *and* The Bat-Poet. The Animal Family *was harder for you, since you couldn't make the pictures direct illustrations—and I'm so grateful to you for working so hard on them and making them so beautiful. It's hard for me to pick the ones I like best—the ones of the moonlight on the sea and the dustjacket itself are my favorites, almost. It will make so much difference to readers having your decorations rather than a book with drawings—the book will feel rich and full to them in a way it couldn't possibly without what you've done. They really are some of your most original and profound drawings—not only will readers of* The Animal Family *love them, but there are all the people who get any book you make the drawings for, just because they're your drawings.*

What you say about The Animal Family *makes me feel awfully good. I like to have it a good book, and when people like you like it as you do, people with so much understanding and remembrance of childhood, much magic and imagination of their own—it makes me hope that book really is what it ought to be.*

We'll be coming up to New York later in the fall, and we'd love to have dinner with you and Michael and go to the opera with you, just the way we have in the past. Seeing you as we have has meant a lot to us.

I'm hoping to write a new children's story now that I'm well, and to have the marvellous thrill of seeing your illustrations for it.

Affectionately,
Randall

About the new book Randall wrote Sendak:

> it's a sort of dream book. All in the present tense [and] is named *Fly by Night*. . . . I'm writing an owl's bedtime story in *terza rima* [for it]. It's more or less the climax of the book . . . It will be so easy to illustrate that I've laughed over the thought again and again. . . . Paragraph by paragraph it divides into pictures, and pictures in your own style.

Little did we know that Randall would never see the pictures he had envisioned. Little did anyone suspect that Maurice himself would be so affected by Randall's untimely death that he was totally unable to create his illustrations for years; and while Randall left a finished text behind in 1965, the illustrated published book was not available until 1976.

Speaking of this in 1998 and how he cherished Randall, Maurice said:

> He was not a grown-up in the conventional sense . . . [And] working with [him] was the most charmed experience of my career. His death was calamitous—is calamitous. I remember the evening I heard about it in exactly the same way one remembers hearing about Pearl Harbor. So, yes, that last book was one of the hardest things I've ever done. I needed the man.

VIII

Douglas Bush writes, "We live in a time when poetry is on the way out." Now poetry—if by poetry we mean what Frost and Dostoevsky and Freud and Ingmar Bergman share—isn't on the way out, unless humanity is on the way out. [If that is so] when poetry "goes out of place it is not the first to go, nor the second or third to go,/It waits for all the rest to go, it is the last."

RANDALL JARRELL

Twelve Years Old: California

RICHARD SEELY

The Lost World

"TO EVERY THING THERE IS A SEASON AND A TIME . . . ,"
that is, a time to plant and a time to pluck, as Ecclesiastes pro-
claimed. I, too, strongly believe in timing; believe that certain
apt and right and *limited times* exist in our lives when certain
significant action is ready for the taking.

For Randall 1963 was the time to write his long, autobiographical poem, *The Lost World.* A year earlier Randall was in a
poetry block and a year later in a depression, but in the
spring of 1963 he had a rush of creativity when he was pluck-
ing poems out of the air, literally, from a conversation over-
heard at the Plaza, from the neighbor's wash on the line,
from the bats on our porch. At just this time Randall's
mother sent him an old Christmas card box from the twenties containing the letters he had written her from California
when he was twelve.

In the mornings after that while Randall wrote, in the Morris chair, in our second living room, I'd look in as I passed and
see that box on his knees and his head tipped down studying a
letter spread out on his palm. On the blue-lined pages he had
torn from a tablet so long ago were his "Mama" and "Pop"
grandparents and his loved "Dandeen" great-grandmother
and their world out West in the house by the tree house on the
street off Sunset Boulevard in 1926.

"Dear Mother," the letters began, and usually ended with, "Tha's all. Lovingly, Randall." Within those boundaries Randall had packed the daily details of nearly a year of his boyhood. Choosing from these and other details from memory, he found the bricks and mortar to rebuild that world in four poems.

The boy in the letters went to the Methodist Sunday School and listened to "grand opera" on his crystal set. He also wrote, "Pop and I explored the Four-Square Gospel Tabernacle. It sure is nice." And, "Dandeen and I rode the double-decker bus like in New York and went up to the new LA library that cost two million dollars. I got an upstairs card." And, "Mama and I went to the County museum. I'd like to spend a whole day there sometime. I liked the animals and bones and weapons best." On the homey side, he reported, "We sure had a good dinner yesterday. We had chicken and squash and potatoes and biscuits and cornbread and lots of other things and peach homemade ice cream and cake. I ate six saucers of ice cream and five pieces of cake." And, in answer to something his mother was checking on, he said, "Of course, I don't bite my fingernails anymore. I'd stopped that when I left Nashville. Why there's not a fingernail on my hand that's even broken."

In "Children's Arms," Part I of *The Lost World*, the fantastic and transporting first scene was evoked by this paragraph:

I saw a picture show being made last Monday and Sunday night. They made it in a big concrete bowl and they had dogs and Eskimos and igloos and icebergs and snow in it. They had a snowstorm. They threw Christmas tree stuff in front of an airplane propeller and it looked like a blizzard.

Further along in that poem is Randall's rapt account of attending *The Admirable Crichton*, beginning "In the black auditorium, my heart at ease" and continuing through sixteen lyrical lines that sprang from these plain ones: "I saw the Senior Play this afternoon. It is the big event of the term and it certainly was good. It had four acts and lasted two hours and a quarter."

In Part II, "A Night with Lions," Randall recalls an idyllic overnight visit with his *"young, tall, brown aunt,"* who let him lie beside her while her friend's lion *"roars his slow comfortable roars,"* and she talked to the boy as a *"grown-up"* about Jurgen and Rupert Hughes, and he thought *"as a child* thinks: *'You're my real friend.'"* What a marvel that this elegant translation could be made from these few flat facts to his mother. "I went to Bettie's and had a grand time. The new cubs sure are cute. They're not any bigger than a cat. I played with Tawny. He wants everybody to play with him."

In Part III, "A Street off Sunset," when the boy is *"forced out of life into bed,"* Randall is paraphrasing a description from a letter that said, "Instead of giving me the little back room, Mama has given me the bedroom between Dandeen's and her own. It sure is fixed up nice. There are some headphones on the bed so that I can listen to music at night in bed."

This is not to say that these rudimentary phrases add to our "understanding" of the poems; but, along with verifying the autobiography, they prove once again with what sow's ears an *artist* can make what the world says one cannot—a silk purse.

Evident in poems I and II are the boy's love for those who care for him and his gladness in their everyday life: and this, too, is borne out in letters. But in poem III Randall's preadolescent anxiety over the wringing of the chicken's neck, the blowing up of the world, and Mama's power over his rabbit's life are not even hinted at to his mother. It took time—thirty-seven years—to reckon with that and to be able to admit his misery at having to return to Nashville. Tucked into the poems are Randall's belated apologies to Dandeen and his aunt for never writing them and for trying to excuse this "*Because . . . I was a child, I missed them so. But justifying/Hurts too . . .*" At last III ends happily with the trusting boy on the steps by Pop "*at the end of [their] good day.*"

This aesthetic ending seemed technically correct and was truthful as well, for he had often experienced it with Pop, and Randall put away the letters and turned to other poems suddenly crying to be written. To his surprise and dismay he could not find one that suited him for *ending* the book, and after thinking about this for some time he wrote "Thinking of the Lost World." Of course it is a limb of the trilogy, but he did not designate it IV, as he wanted to keep it separate, as it is in time. When, like Dumas—whom Randall refers to in "Thinking of the Lost World"—he, too, brings his D'Artagnan back decades later for a second bow, Randall bestows a second ending to *The Lost World*. Revisiting the scenes of his boyhood, the poet settled in North Carolina and married, finds "*The Land of Sunshine . . . a gray mist now . . . [his] arrows are lost or broken . . . and the planks of the tree house are all firewood/Burned long ago.*" In accepting and forgiving these changes in happiness even, "Thinking of the Lost World" provides the real conclusion to his boyhood year in California.

Had Randall been a few years younger when that hoard of letters arrived, *The Lost World* might well have had more of the bitterness and hopelessness of "Hope" in it. But no: he was fifty and the old break was mended. A few years earlier, too, when Randall was still undaunted by *length*, *The Lost World* might have achieved Homeric proportions. But no: it was a time to tighten, to take the discipline of terza rima, and to have done with longueurs. In this time of change Randall advanced to new ground—ground not hallowed by Auden and Rilke—*his* ground, his personal property. How often that spring he'd say to the family, "Just think, Old Ramble has written four-fifths of a book."

All but three of the other poems in *The Lost World* were written over the three-year period before it was published. The new conversational directness of these poems, their contemporary settings, and the increased personal involvement distinguish them from the three earlier, more rationalized and discursive, poems: "A Hunt in the Black Forest" (1947), "Woman" (1953–63), and "Hope" (1958).

In all probability "A Hunt in the Black Forest" marked the close of Randall's season with the Märchen on the Rhine. I doubt we would have seen another poem of that kind in his next book. Not so, with "Woman": Randall's affinity for her goes as far back as "Lady Bates" and "The Face." In "A Girl in a Library," "Seele im Raum," and "Cinderella" he experimented with portraying women as the observed or the self-observer, and in "The Woman at the Washington Zoo" he perfected the first-person dramatic monologue he carried over to "Next Day" and "The Lost Children" and his later women-poems. When Cal praised Randall in a letter for his "great parade of women," Randall decided to withhold three recent poems— "A Man Meets a Woman on the Street," "Gleaning," and "The Player Piano"—for his next book, a book, he wrote his editor, Michael di Capua, that was to be called *Women*.

Actually, Randall was in thrall to the feminine mystique, and life gave him a habitat rich with a mother, grandmother, and great-grandmother; an aunt and aunt-figures; two wives, two mothers-in-law, and two stepdaughters. Furthermore, as Freud had said, "There *are* no accidents," and it was no accident that Randall taught at what was formerly the *Woman's* College of the University of North Carolina, now coed. How could he

not have written about women? Oh, but he wanted to, and details and traits from all of us glow and sparkle in his female
composites. Randall *liked* women, and he meant it when he
wrote:

> *In the beginning*
> *There was a baby boy that loved its mother,*
> *There was a baby girl that loved its mother,*
> *The boy grew up and got to love a woman,*
> *The girl grew up and had to love a man.*

> *"Woman,"*
> *The Complete Poems*

Was I a source for any of this? Whether it was Venus or
Medusa, the Marschallin or Mother Death, many of Randall's
poems—but not all—have shown me something of what I
was for him and something of what I was not. In the Berg Col-
lection of the New York Public Library there are more manuscript pages for "Woman" than for any other poem of
Randall's. This is because from 1953 onward he kept adding
everything he could think of and then taking out everything
he could spare. He was always enthusiastic about "Woman,"
and in a version he sent Cal he wrote, "Nobody will have to
read Pope's *Characters of Women* anymore."

I could be wrong, but I don't see myself in the wife in "A
Well-to-Do Invalid," or the blond female in "Three Bills," or
the eccentric in "The One Who Was Different." I believe that
"Next Day" is a persona mask for Randall's own melancholy
over growing older and reacting to not one, but two friends'
funerals in one month. While we often talked about our aging
and "the meaning of life," this was more troubling to him than

to me. I do concede to being the mother in "The Lost Children." The dream of the sunny house in the poem was a dream I told him, and those were my photographs of my daughters; my feelings about my empty heart that had "*lost a child, but gained a friend.*"

"Write it all out for me, old pet, just the way you've said it. Please?" he insisted: and I did. Then, as was his custom when ideas for poems were gathering, he took up residence in an armchair with a fresh notebook on his knees and an insulating layer of stereo sound around him—Chopin at that time. I remember sitting on the porch with Elfie in the pretty weather for several days, sewing the summer covers for some cushions. In due course I heard him calling out, "Where's my magic girl?" And when he found me, he said, "Put down that wicked needle, child. Hold out your hand and shut your eyes. And, I'll-give-you-something-to-make-you-wise." It was my notes smoothed into meter and narrative, and transformed into the poem "The Lost Children."

"The Lost Children" was, as Randall said, "sure fire" with audiences. After a reading I would see girl students come up to him with eyes moist from daughter-remorse, wanting a copy of That Poem to send to their mothers, and middle-aged women, smiling gamely, wanting That Poem to send to their daughters. How uncannily *right* for them, and for Randall, was this poem about a dream, a woman, childhood, and loss. He often shook his head over this, saying in a mystified tone, "Believe me, this is *echte* Jarrell."

A complete departure from *echte* Jarrell are the three nature poems: "The Mockingbird," "The Bird of Night," and "Bats"— yet they triggered this book. During Randall's poetry block after 1958, he mainly wrote criticism and translated Goethe's *Faust: Part I*; and in early 1962 he was hospitalized for hepatitis and could not write at all. Then enter Michael di Capua. Talking about this in a radio interview, Randall said:

> I got back to writing poems again in a funny way. . . . I was extremely sick. For the first time in my life, I was bored with reading. And Michael di Capua, who was a children's editor at Macmillan then, happened to write to me and he knew my Märchen poetry and so on, and he said, "Wouldn't you like to translate some Grimms' tales and write an introduction," and I said, "I sure would." And so I did different ones. I was gradually getting to feel better and I got to know Michael pretty well, and he said, "Why don't you write some children's stories? You're always writing about children in dreams and fairy tales." And so, while I was convalescing that spring, I would sit out in the yard where my wife was gardening and write. I wrote a little story for children named *The Gingerbread Rabbit*. And after that I started writing another book half for children, half for grown-ups, named *The Bat-Poet*. And that felt just like a regular book to me. You know how it is, you work all the time on it, and stay awake at night. And wake up in the middle of the night and—Anyway, by good luck, we had some bats on our front porch, and I imagined a bat that would not *write* poems, but make them up. And anyway, *I had to make up poems for him*—Oh, but in fact, I wanted to make up poems for him. And a couple of the poems were pretty much like grown-up poems. Anyway, *The New Yorker* printed them. Well, by then I was writing poems again and so I wrote two-thirds of *The Lost World*, and so you can see how extremely grateful I am to Mr. di Capua.

Since *The Bat-Poet* prompted *The Animal Family* and *Fly by Night* and the bat's poems unblocked his own, Randall was overjoyed the day we thought of dedicating *The Lost World* "To Michael di Capua from Randall and Mary."

The Lost Children

Two little girls, one fair, one dark,
One alive, one dead, are running hand in hand
Through a sunny house. The two are dressed
In red and white gingham, with puffed sleeves and sashes.
They run away from me . . . But I am happy;
When I wake I feel no sadness, only delight.
I've seen them again, and I am comforted
That, somewhere, they still are.

It is strange
To carry inside you someone else's body;
To know it before it's born;
To see at last that it's a boy or girl, and perfect;
To bathe it and dress it; to watch it
Nurse at your breast, till you almost know it
Better than you know yourself—better than it knows itself.
You own it as you made it.
You are the authority upon it.

But as the child learns
To take care of herself, you know her less.
Her accidents, adventures are her own,
You lose track of them. Still, you know more
About her than anyone except her.

Little by little the child in her dies.
You say, "I have lost a child, but gained a friend."
You feel yourself gradually discarded.
She argues with you or ignores you

Or is kind to you. She who begged to follow you
Anywhere, just so long as it was you,
Finds follow the leader no more fun.
She makes few demands; you are grateful for the few.

The young person who writes once a week
Is the authority upon herself
She sits in my living room and shows her husband
My albums of her as a child. He enjoys them
And makes fun of them. I look too
And I realize the girl in the matching blue
Mother-and-daughter dress, the fair one carrying
The tin lunch box with the half-pint thermos bottle
Or training her pet duck to go down the slide
Is lost just as the dark one, who is dead, is lost.
But the world in which the two wear their flared coats
And the hats that match, exists so uncannily
That, after I've seen its pictures for an hour,
I believe in it: the bandage coming loose
One has in the picture of the other's birthday,
The castles they are building, at the beach for asthma.
I look at them and all the old sure knowledge
Floods over me, when I put the album down
I keep saying inside: "I did know those children.
I braided those braids. I was driving the car
The day that she stepped in the can of grease
We were taking to the butcher for our ration points.
I know those children. I know all about them.
Where are they?"

I stare at her and try to see some sign
Of the child she was. I can't believe there isn't any.
I tell her foolishly, pointing at the picture,
That I keep wondering where she is.
She tells me, "Here I am."
                    Yes, and the other
Isn't dead, but has everlasting life . . .

The girl from next door, the borrowed child,
Said to me the other day, "You like children so much,
Don't you want to have some of your own?"
I couldn't believe that she could say it.
I thought: "Surely you can look at me and see them."

When I see them in my dreams I feel such joy.
If I could dream of them every night!

When I think of my dream of the little girls
It's as if we were playing hide-and-seek.
The dark one
Looks at me longingly, and disappears;
The fair one stays in sight, just out of reach
No matter where I reach. I am tired
As a mother who's played all day, some rainy day.
I don't want to play it any more, I don't want to,
But the child keeps on playing, so I play.

The Complete Poems

On the one hand, *The Lost World* poems have been dubbed "sentimental," and on the other, "embracing a wide human involvement." They are seen by some as a "diminishment"; others remark on their "wonderful resilience." Poet Dave Smith writes that "Jarrell wanted things pure and wanted truth expressed. Yet he knew the meanness of the human soul and he was passionate. [This] contention of ruthless intelligence and the justice of aroused compassion mark his poems and keep them forever near me. . . . If you have business to do with poetry and people you have to go to him." Suzanne Ferguson, the author of *The Poetry of Randall Jarrell*, believes *The Lost World* "affirms the worth of individual human efforts to understand and change themselves, and, if possible, the world. Though he sets [his] efforts in his own particular time and place, their impulse is timeless, universal, and transcendent. The world of Randall Jarrell is a world that does not get lost."

IX

*The one incontestable truth about love
is that it is a mystery
and all that is written about it
is a series of questions
that have remained unanswered.*

Anton Chekhov

The Group of Two

TED RUSSELL

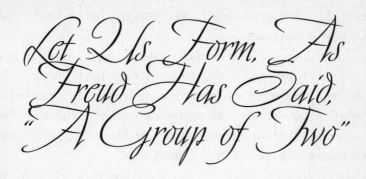

Let Us Form, As Freud Has Said, "A Group of Two"

In July 1951 on the never-to-be-forgotten day Randall and I met, we were both thirty-seven. Randall had flown to Boulder from the Wianno Club summer tennis tournaments at Cape Cod. My mother and I had train traveled from California on the Atcheson, Topeka and Sante Fe because she was a stockholder.

On arrival at Clark Hall on the University of Colorado campus for the Rocky Mountain Writers Conference, my mother retired to our room for her nap, and I registered (at a tuitional cost of twenty-five dollars each) for the lectures on How to Make a Long Story Short. How to Make a Short Story into a Novel. How to Write for Magazines. For Children. For Hollywood.

After that I joined the Writers Orientation Tour of the campus and fell in step with a tall, slim, young man in a Basque beret and handmade leather sandals. He said his name was Dee Snodgrass and that he was there for poetry, "For Randall Jarrell." Had I heard of "The Death of the Ball Turret Gunner"? I hadn't. Meanwhile, the unbeknownst-to-me poet I would meet that evening and marry the next year was playing singles with the university's ranking player as had been prearranged for him at his request.

After dinner and announcements in the dining hall, the conferees adjourned to the student lounge for a "mixer." My mother was already mixing with an older writer named Marjorie Dodson, who providentially was from St. Louis and who had attended Mary Institute a few years after my mother, and they were into their "Did you ever know?"

From across the lounge came a tinkling little three-finger tune and standing over a baby grand was a deeply bronzed young man with dark curly hair, a Ronald Colman mustache, and wearing a burnt orange linen shirt. He caught my eye—or I caught his—and I drew near. And he said, "Do you play?"

"Not really. Do you?"

"Only this," he said blithely. "It's my 'little phrase,' like Vinteuil's. You know? In Proust? My name is Randall."

"And mine is Mary."

He left off playing his little phrase and said suavely, "We don't want to stick around for this dopey mixer, do we? Why don't we set out for the great world?"

As soon as we were clear of the student lounge, Randall stopped us at a drink machine and bought us each a Coca-Cola to drink on top of dinner.

This new friend was as new on that campus as I was. In fact, newer, because he had skipped the Orientation Tour to play tennis. Nevertheless, what began as an aimless wander past the bandstand—while talking the whole time—magically led us up the steps of the library. So like him. Randall never entered a library he didn't feel at home in. In no time, he had me settled in the reading room with a volume of Chekhov's stories opened at "Rothschild's Fiddle" and was off to the stacks to find books for his classes the next day. Before our evening was out, Randall advised me in a kindly, professorial tone, "Oh, Mary, wash out those dumb classes you've signed up for and just sit in on mine. You'll like it. I guarantee."

For a fraction of a second, one hundred and twenty-five dollars flashed on in my brain. What if they wouldn't refund it? But I took the risk. And they did. And our Group of Two was forming.

I entered Randall's life and lifestyle after his stint in the Air Force and his stints at Sarah Lawrence and *The Nation*, and after Peter had wangled him a job in the English Department at the then Woman's College.

I got to stay through Chekhov, Goethe, and Rilke. Past Rene Clair to Ingmar Bergman; past the dancers Danilova and Tanaquil le Clèrq; from sopranos Lotte Lehmann to Schwarzkopf; and all through those elegant stylists in their fields: Phil Hill, Pancho Gonzales, and Ted Williams, the Red Sox hitter; Johnny Unitas and Fischer-Dieskau, the superb German baritone. And, oh *yes*, JFK.

Randall's Enthusiasms put a spell on him and on you, too. I watched many a sensible person who met him during an Enthusiasm come away sure in the knowledge that Randall's main side interest was graphology. The next person was sure it was gear ratios. Those were only the transitory Enthusiasms Randall paid attention to "As a Novelist," often finding a niche for them in poetry or criticism and always enjoying them for the time being. With *real* subjects—anthropology, Renaissance painting, psychoanalysis, romantic composers—he dwelt in them and they dwelt in him.

Randall was an active spectator. In fact, if he was merely sitting idly by, *watching* a performance, I could be sure he was about to say, "Let's wash this out, von S." Anything worth stay-

ing for made him twist and untwist his legs, clasp and unclasp
my hand; jerk, hum, and sweat with empathy until he almost
"rang" like the line of Rilke's he translated: *Whenever I saw something that could ring, I rang.* On the way home he would say, "Oh, I
ought to write Ingmar Bergman a letter." Or Groucho Marx. Or
Jack Benny. He'd wanted to suggest to Balanchine that he
choreograph a ballet for Schonberg's "Transfigured Night,"
but never did. In the case of Sviatoslav Richter, Randall went
so far as to have a Gucci handbag delivered to Mrs. Richter, but
while he could discourse at length to me about the great pi-
anist, he couldn't express himself on the card with more than
his signature.

I didn't read *Road and Track*; Randall read it to me after he'd
read it himself, then he'd read it himself again. When the new
ones came, the old one went under the bed with old *Schwann*s
and *L. L. Bean*s and *High Fidelity*s.

In *Road and Track* Randall got a world peopled and costumed
and behaving as imaginatively as he could have made it himself. Best of all, he hadn't made it himself. *Road and Track* came
nearer than anything else to satisfying his leitmotiv of yearning, in certain poems, for "people from another planet." On
our deck in Laguna, looking past the Seal Rocks to Catalina Island, he'd dream over those drawings of dreams—The All-
Possible Car. He *believed* in *Road and Track*'s theocracy of old, holy
motors whose Dalai Lama was the adorable Bugatti. And he'd
marvel at the exotic vernacular critics used to review drivers
and drivers used to review cars. "Baby *doll*! What prose!" he'd
say. "Listen to this."

There in that magazine, among dates of rallies and *concourse
d' elegance* and races, man found the dates of consequence to
man. "Just think, pet cat," the bridegroom said when he'd
found our wedding date. "We're probably the only people in

the world getting married on the eighth so we can be at Madera on the tenth."

It was a fine, bright, smog-free day in Pasadena and clear enough to see Mount Wilson Observatory in the distant Sierras beyond my mother's garden. In fact, the *clear* day without smog was the main topic of the wedding guests, and Randall muttered to me, "One touch of weather makes the whole world kin." It was *Gemütlich* there, with my friends of twenty years perpetuating us in Kodachrome. Randall's eyes fastened on mine above our champagne as we toasted *ewig Freundschaft* to each other, *ewig, ewig*. Red-haired Helen Dengler's accordion ebbed and flowed in the rooms, and it was night when our last good-byes drifted off into the eucalyptus-Vicks-scented air. Madera. Another world. And two Jarrells perched on a rail fence oblivious of all save ourselves and Phil Hill racing the first C-Jag in this country.

To be married to Randall was to be encapsulated with him. He wanted, and we had, a round-the-clock inseparability. We took three meals a day together, every day. I went to his classes and he went on my errands. I watched him play tennis; he picked out my clothes. Sometimes we were brother and sister "like Wordsworth and Dorothy" and other times we were twins, Randall pretended. "The Bobbsey Twins at the Plaza," he'd say up in our room at the Plaza. And he'd laugh and add from *The Three Sisters*: "The silliest one in the family. That's Masha." He got us trading scarves and gloves the way he and Peter traded hats and jackets and he and his colleague Bob Watson traded ties.

Like the boy in Rilke's "Requiem," Randall "printed" on himself the names of Chekhov and Proust and Goethe. When some guests mistook our small, framed picture of Chekhov for Randall's father, Randall shook his head after they had gone and said, "Is there no limit to what people don't know?" But before we went to bed that night, he took a long look at Chekhov's photograph and a long look at himself in the mirror. "You know what?" he said. "If you blur your eyes . . ." And I said, "It's so, Randall. It's so." The next day he stopped shaving.

Years after, when people said he looked like Renoir's gentleman with the opera glasses in *La Loge* or like Donatello's head of Goliath, he'd ask with innocent joy, "I declare. Do you really think so?" Gradually, with no overt plan, bearded pictures outnumbered the others in every room. Solomon, Odysseus, Constantine, John the Baptist, and *der heilige Hieronymus* became household favorites.

Randall had an affinity for what he thought of as his Other: that One he saw in ponds and photographs and mirrors. He had favorite and unfavorite mirrors but I believe he looked into all he ever saw. After a day in bed with a bad cold, when he had read and written and slept too much, he would ask me to bring him the hand mirror. Unperturbed by my watching, Randall studied his face with discernment, thinking I know not what. Then he'd turn playful and facially pantomime Tsar Nicholas turning into Rasputin and Mephistopheles turning into Faust. Waiting in the Mercedes or the Jaguar for a traffic light to change, Randall seemed to be looking sideways into space, but when I followed his glance, there we would be, *the windowed ones within their windowy world*, in the reflecting plate glass of a storefront. I would wave to the delectable car and the man at the wheel, and to the woman peering over his shoulder, waving back. I'd say, often, "There they are, my Randall." And he would smile at them and say, "America's dream, my mermaid."

Along with our almost daily visit to the university or the

downtown library, we checked the morning and afternoon mail in our post office box. "Who knows, eh?" Randall would say. And yes, there could be fan letters or letters from Peter or Cal. And query letters for speaking engagements, writers' conferences, contests, articles, and reviews. Randall liked them all except the ones with a California postmark. These he stuffed into the glove compartment unopened, saying, "some feeble-minded hippie." He could also tell from the outside, somehow, which letters were "Ball Turret Gunner" permission requests. He'd hand them over to me for reply and we made no comment. It was an awkward moment. I'd write back granting permission for the five-line poem to be used in all kinds of giant anthologies for a fee of twenty-five dollars. It wasn't the fee so much. Twenty-five dollars bought us a restaurant dinner in those days. It was the principle. What stung Randall was the custom of textbook departments epitomizing his two books of acclaimed war poetry and three books of civilian poetry to this *one* space-saving and money-saving five lines. We talked of writers who were known for one novel, one play, one book, and each "Ball Turret Gunner" letter stabbed Randall with the notion he was a one-poem poet.

Two little girls, one fair, one dark—one Alleyne, eleven; one Beatrice, eight; my daughters—came into the marriage with me. To them, Randall was more of a friend, or pet, or affectionate encyclopedia than a father. Kitten came with Randall. He and Kitten had taken walks and naps and trips together for years. They trusted and admired and "went out" to each other. To Randall *every part of Kitten had a "clear, quick, decided look" about it.* That sentence is quoted from a description of the mockingbird in *The Bat-Poet*, but Randall said it first about Kitten. These were Kitten's qualities and Randall wanted them from tennis partners and automobile mechanics and critics. At certain intellec-

tual moments Randall had that look himself, and he and I often caught a glimpse of it when Ted Williams was at bat or
when President Kennedy was fielding questions at a press conference. Simultaneously we would say, "Look. Look. He's looking like Kitten."

Randall said, "Kitten has a general's eye for terrain." And in
any new neighborhood Kitten soon had in his head which
houses he could get under and which trees and roofs would
serve best for ambush and reconnaissance. After some three-
day bivouac (worrisome for us), he'd emerge from a hedge and
Randall would drop anything to run out and pick him up and
hug him and cry in a breaking voice, "Oh, Kitten. *Kitten.* Clever
one to take such good care of yourself."

They played a game based on mutual anticipation and fast
reflexes, very like tennis, that consisted of Randall's flicking
the end of a necktie in provocative ways just out of Kitten's
reach—if he could. They would range over the floors and furniture, high and low, and play several "sets" with the easy skill
of two "clear, quick, decided" intelligences that had never
failed, or been clumsy, at anything they'd ever done.

When Kitten thought Randall had been writing too long,
he sometimes stepped into the middle of the room and made a
long, drawn-out, vexed cry of impatience that brought Randall to his feet. Other times he took a stance directly in front of
the chair and with his ankles together and his tail afloat he'd
stare up at Randall, giving off rays of invitation. Soon I'd hear
the familiar melody, "Little ambassador, are you bored? All
right. We'll play. Come now. We'll laugh and play. Oh, yes. Oh,
yes."

When Randall relaxed in the bathtub after tennis, and read,
Kitten often sat nearby and purred. He could purr until he almost sang. When his heart was especially happy he'd stand

resting his front paws on the edge of the tub and he'd stretch until Randall met him halfway with his bowed head and they'd bump and rub heads. If Randall were lying on the sofa reading, Kitten would lie along the arm of it with his head near Randall's, black on black, and lick Randall's hair with the same patient, purposeful strokes he used to "wash" his own.

Once at a cocktail party on a campus we were visiting, a professor began drawing Randall out on Kitten. Delighted to escape from General Conversation, Randall was animated and voluble and a circle soon formed. At this point, to be funny perhaps, the professor interjected a story he'd heard about Randall giving his meat-ration coupons to the cat during the war. "Why of course!" Randall flashed sparks. "What would you *expect*? He's only a poor cat, and has to eat what he can. People can eat anything. What an absurd remark."

At home anyone holding Kitten on his lap had, as the girls said, "King's X." If the music on the FM "went bad" while Randall was holding him, I'd hear him in the other room, saying "Help! Help! Someone come change the music. I'd do it myself but I'm holding Kitten."

On many clear nights, back from the library, we'd put the car in the garage and be looking up to find Orion's belt and sword when a downy, warm, almost invisible presence entwined itself in and out between our ankles, leaning lightly. "Isn't that *polite*?" Randall said. "To want to come and meet us!" One night he didn't come. Kitten was hit at the side of the road by a car. Like Randall, one blow on his skull killed him instantly. Again like Randall, the beautiful eyes and face, and the graceful body, were not hurt. Alone together we buried him in the dark under a deodar tree. When I laid some fern leaves down in his grave first, Randall *thanked* me. Then we curled him in a circle, "like a little fox," and covered him over. At class the next day, Randall could scarcely teach and we decided to drive to Charleston for a few days. When we came back, people spoke so kindly to Randall. Alleyne and Beatrice were so loving; and I did all I could to comfort him, but in the end he had to suffer by himself the zigzag work of mourning, with the guilt and the longed-for dreams, the dreaded fading and the reluctant giving up.

Something we never got to the end of talking about was California in the Old Days. We so nearly must have met back then that we thought we *must* have met. Randall and I had lived in Long Beach at the same time. We knew each other's houses. Mr. Jarrell worked for a photographer named Richard Seely. Randall's father called on prospects for the Seely Studio and once someone, *someone!* called at my father's office and arrangements were made for a series of photographs to be taken of me as a present for my mother. They are gently lighted, altogether winning studies, as are the ones he made of Randall at that time, and of Mr. Jarrell, too, and all signed Richard Seely in India ink.

The photographs of Randall's father are of a fair, gray-eyed, aquiline-featured young man not even thirty. He is in his double-breasted vest and wearing a wide-brimmed fedora; he looks like Wallace Reid. And I heard with a pang how sharply he'd been scolded for buying the pretty, striped shirt he had on, when they were short of money.

We talked a lot about the reception room my father shared with another doctor, Dr. von W., who was treating Mrs. Jarrell for that *recurrent/Scene from my childhood,/A scene called Mother Has Fainted.* I sat in that room after school sometimes, waiting for my father to see his last patients and give me a ride home. Randall went there too—to wait for his mother. I clearly remember chatting with the receptionist in front of the others and feeling important because she knew me. And I faintly recall that some schoolboy in sneakers was often sitting by the window, absorbed in his magazine.

In Nashville, Randall said, he was "covered with relatives." The Campbells (pronounced Cam'll) were an intimate, dominating family of strong wills, the whole of which was not as formidable as its parts. None of them was a listener or a relaxed person, but each on his own level was effective and, as Randall's mother said, "left tracks." She was the poor and pretty one who "had to work" and Uncle Howell was her mainstay. In Campbell minds, Randall was expected to Be a Little Man and to aim toward supporting his mother, which, unhappily for Randall, Uncle Howell had done for his mother at an early age.

"They had real gifts for finding me the most *awful* jobs," Randall said. "I wouldn't have minded delivering papers so much—though it was hellish—if I could have hired somebody to do the collecting. The people were so *bad*. They wouldn't pay, and they told lies. And I had to keep going back." They made him sell Christmas seals and ribbons from house to house, and Randall said, "*Imagine*, pestering people like that in their houses. Wasn't that a wicked thing to make a child do?"

When the sculptors for the Nashville Parthenon invited Randall to model for Ganymede, cup-bearer to the gods, they were enchanted with this child who told them myths of the gods while he posed. He soon had the run of their studio, spent whole days with them. When they finished the pediments, though, they had to go back to whatever planet they'd come from and Randall was left desolate. Long afterward his mother revealed that the sculptors had asked to adopt him, but knowing how attached to them he was she hadn't dared tell him. "She was right," Randall said bitterly. "I'd have gone with them like *that*. . . . My mother is a disaster."

He went on growing up in the Carnegie Library and at the backboard of the public tennis courts and writing in his room with the door locked. In high school he played on the tennis team, edited the literary magazine, and headed the honor roll. It was the depression and at his high school graduation Uncle Howl offered him a job in the candy company, provided he learned bookkeeping and shorthand at the secretarial school. Randall sat through the classes but said he couldn't make himself pay attention and would draw airplanes or write poetry or put his head down on the desk. Eventually, Uncle Howl sent him to Vanderbilt as a day-student—the first one in his family to go to college—and Randall was grateful.

He was still poor. He told me about playing tennis with one nickel in his pocket and how it distracted him from his serve—trying to decide whether to keep the nickel for carfare home or spend it on a bottle of Tru-Ade to pour over his head. When we were married he was temporarily hard up again. Strewn through the handwritten pages of *Pictures* are little sums like $23.80 plus $41.50 with $14.95 taken away. He would sigh, "When I get through this, all I want is enough money not to have to think about it." Even then, Randall was above thinking about change, and pennies poured out of his pockets into the sofa and car and onto the bear-fur rug by our bed. Alleyne and Beatrice called these "Randall's oil wells" and he let them keep their findings. "A penny is more trouble than it's worth is my motto," he said. Change got crammed into any pocket and weeks later he'd come into a room all smiles and say, "Guess how much money I found in assorted jackets this morning?" Randall never had a savings account, only a spending account where his royalties and honorariums and salaries were transubstantiated into opera, the house in Montecito, the antiques, and a hand-carved life-size swan we bought with his honorarium from Johns Hopkins. After a childhood in which a good hand-kerchief had been a miserable present, Randall bought his at Sears and used them as a grown man, for tearing into strips when he wanted to tie something, or wiping the grease off the wire wheels of the Jaguar.

The last time we were in Europe we saw a spotted fur hat with a Garbo-dipped brim that Randall said was "right up your alley." I thought so, too, but I was ashamed to spend a hundred dollars for a hat. And I didn't want him paying for it because he'd already bought me a present and it was my turn to find him a present. I hated to take off the hat and kept looking at myself in the glass, not knowing what to do. "How much have you spent on hats in the last ten years?" Randall asked me. "Nothing!" I said. Californians didn't buy hats. "Okay," he went on. "Prorate it. If you allow yourself ten dollars a year for the past ten years, does it seem so bad?" I laughed, took out my checkbook, and said, "Fur hat, will you marry me?"

We forgot about this until Randall's nervous breakdown overtook us. When, in Rilke's words, "*rain starred the stream*" and we saw "*the naked tree Trouble.*" Then we talked over what *real* trouble we'd had in our years together. It seemed so little to both of us that we said if we'd prorate our bad time just then against the rest we'd still come out way ahead.

After our two years in Washington we bought a rustic house on a red clay road in a small forest of pines and hardwoods. "The Bobbsey Twins in Their Hunting Lodge," Randall called us. I think this was the time that, just to be mischievous, I said to him about the twin-ing, "How can that be, Ramble? I'm four days *older* than you." "Only chronologically," he sang to me over his sci-fi magazine. We fed and read about birds and bought and read about trees. Randall learned how to put up hammock hooks and he put up a dozen. It was his idea to surround the house with rooted ivy plants in hopes they would take the place as the neighbors warned, saying "The first year it sleeps. The second year it creeps. The third year it leaps." And yes, they were right. When ivy stayed across the window screens and, to Randall's delight, sent tendrils indoors, the neighbors were horrified and triumphant and asked him what he was going to do *now*? Randall said, serenely, "Let 'em. They won't harm me if I don't harm them is this house's motto."

Cal's letters in this period were full of the grim realities of the bomb and mass death, and they stuck in Randall's mind and set him to brooding. "But Cal is right," he said and wouldn't be comforted. "What an age to be part of!" He was gloomy that spring until he translated a sentence of Luther's that seemed to ward it off: *And even if the world should end tomorrow I still would plant my little apple-tree.* He quoted this to Cal and to classes, and put it in scansion in the front of his book *A Sad Heart at the Supermarket*. Then he bought us a six-tree apple orchard, some hollies, firs, golden willows, ginkgoes, a magnolia for me, and a birch for "good, sweet Chekhov."

Alleyne had married her med student and Beatrice was off at college. Randall and I were having meals at all hours and playing *Tristan und Isolde* in the moonlight. In our dark living room far across from us, the tuner's green lights made it Daisy's dock that Gatsby watched, we said; and the amplifier's many-heighted tubes glowed for us "like a little city."

On Sundays we had pasties with Pilsner Urquell and watched the National Football League on television. Beside quarterbacking plays, Randall was continually appreciating scenes of the crowd, half in light and half in shadow. Or of half stadium and half turf with the athletes in combat on the limed lines of the grid. "Wouldn't that make a painting!" he'd say. "Oh, if only I could *paint*."

Randall didn't join things, unless you count Phi Beta Kappa, the National Institute of Arts and Letters, and the Army. If he "had a hard time knowing what to do at parties," it was even worse at meetings. He attended when he had to, protesting, "The trouble is, there's nothing to *do*." And when he got home I'd ask what happened and he'd say blankly, "Happened? Well, I cleaned out my billfold." Once, at an *American Scholar Editorial* meeting, he sketched all evening and brought home a speaking likeness of Margaret Mead. At graduations he passed notes to the other professors and made funny signals to me out in the audience. I never joined things either, not even church then, and when people asked if I belonged to some league or some club, I'd have to say truthfully, "I just belong to Randall." When Chancellor Otis Singletary asked us for a mailing list of our friends for the University Tribute at Chapel Hill, Randall's brow got wrinkly. Finally he said, "Otis, we know more chipmunks than we know people." In spite of this the university did find two hundred guests to invite for dinner and two thousand came for the Tribute Lecture given by Red Warren. As a preamble Red chose to remember Randall thirty years before at Vanderbilt, as a student in his sophomore survey class, "Beowulf to Hardy." Red began:

> And among my students in that sophomore section was a tall skinny young man, just getting acquainted with the art of the razor, an art he has now neglected. He was a freshman, but since he had read everything, and remembered everything, he clearly didn't belong in Freshman English. In fact, he didn't belong anywhere. But they put him in the Sophomore Survey, Section A.

He was a very gentle and polite young man, though sometimes, in the innocence of youth, overwhelmed by the spectacle of human dumbness. At such times he might cover his face and moan aloud. But such pity and despair might quickly pass as some flash of intelligence and perception, however rudimentary, appeared before him. Then his face would light up with generous appreciation and deep impersonal joy.

It quickly became apparent that the members of the Sophomore Survey, Section A, did not care much about the marks on their papers. They watched the face of the tall, skinny young man to see how they were doing. So did the instructor, and on days when the hour passed with not once that certain expression of glazed pity on the skinny young man's face, the instructor took a deep breath and hurried out to light a cigarette. The instructor learned a lot in Sophomore Survey. He had to.

The years 1962 and 1963 brought us downs and ups. A big down was Randall's hepatitis but a pleasant up was Ted Weiss's nomination for Randall's Honorary Degree from Bard College.

It was another phase of more prose than poetry, but every year seemed like that to *him*. But happily, the story in *The Bat-Poet* demanded four nature poems—themselves a new genre—which providentially opened the poetry valve another time. Randall's successful nomination of Peter for the National Institute of Arts and Letters was almost as cheering as Peter's decision to leave Ohio State and "if you can wangle it for us" return to UNCG and teach. "A nothing!" Randall told him and jubilantly announced to our mutual friends, "Just think, old Peter Bell got me *my* job here and now I'm getting him *his* job here." In May 1962 an invitation came from Librarian Quincy Mumford in Washington for Randall to give the opening address at the first National Poetry Festival, to be held at the Library of Congress in October. Summer in Monteagle with the Taylors was just what he needed to hone his skills as critic for the topic of the address: "Fifty Years of American Poetry."

Louis Untermeyer, the resident Consultant-Laureate, had instigated the festival; the Bollingen Foundation funded it; and it was open to all the poets in America. It began with a reception in one of the exhibition halls, where the poet's manuscripts were on display under glass. Among the guests who greeted and did not greet Randall, Ransom was fond, Shapiro was sunny, newly met Ogden Nash a friend at first sight. Wily Frost was playing the Grand Old Man, Wilbur was mannerly, and Rukeyser and Eberhart ungrudging. Untermeyer's two former wives, Jean Starr and Virginia Moore, both poets, were sociable but the Tates were aloof. Snodgrass—a Pulitzer by then—was merrily enjoying everything while Berryman and Delmore Schwartz—an even merrier pair—brought John Barleycorn along for company in place of their wives.

Leonie Adams and Louise Bogan were civil but Oscar Williams refused contact with Randall by hand, word, or eye. Elizabeth Bishop, in Brazil, had declined because of the distance; Lowell was in the Psychiatric Institute for Living; Roethke and William Carlos Williams were in failing health. Warren was hospitalized for tests and Sandburg (rumor had it) would not share the stage with Frost. Viereck had missed his flight from Lisbon, someone said. Absent Aiken, MacLeish, Patchen, and Winters were unaccounted for: hardly Randall's fan club.

There were to be three days of panel discussions in the mornings, poetry readings in the afternoons, and a formal lecture each evening. Welcome from President Kennedy was conveyed by the Special Consultant on the Arts, August

Heckscher, and an invitation for the ladies to have sherry with Mrs. Kennedy. Buffet suppers for everyone were planned at Senator Udall's and former Attorney General Biddle's.

This being Camelot time the poets' wives were smartly attired and there were fur coats among the cloth. Fur-loving Randall, unable to restrain himself, lightly touched one finger to the then Mrs. Untermeyer's sleeve and murmured mostly to himself, "What a splendid mink coat." Mrs. Untermeyer beamed happily and said, "There are three splendid mink coats here and Louie bought them all."

At a luncheon one day Randall was seated between a glowering Blackmur, saying nothing, and a gruffy condescending Rexroth. When Randall idly asked him, "How did you happen to learn Chinese?" Rexroth snapped, "Oh, Jarrell, everyone knows Chinese." The whole table was struck dumb until Ogden Nash—soberly studying his menu—broke up the logjam, crying out, "Oh-ho! Back-fin lump. Back-fin LUMP. Oh, I mustn't forget *BACK-FIN LUMP.*"

To cover fifty years of American poetry in one hour Randall singled out fifty-seven poets to mention. Who got the most space was significant, who got grouped with whom was significant, and who placed at the finish line was to die for. As Berryman said, "All of the poets sat on the edge of their seats while Jarrell put together the jigsaw of modern poetry in front of our eyes." All, that is, except Bogan, who retired early and missed hearing herself paired with Adams as "a poet in the Elinor Wylie and Edna St. Vincent Millay tradition of feminine verse." Absent Bishop and Marianne Moore missed hearing themselves defined as "poets in a very different tradition who seem to me the best women poets since Emily Dickinson. And," he continued, "an extraordinarily live, powerful, and original poet, Eleanor Taylor, is a fitting companion of theirs." Concluding, Randall gave Wilbur two hundred and thirty words, Shapiro, two hundred and fifty, and Lowell seven hundred. About Cal, Randall said in part:

> The subject matter and peculiar circumstances of Lowell's best work, for instance, "Falling Asleep over the Aeneid," "For the Union Dead," "Mother Marie Therese," "Ford Madox Ford," "Skunk Hour," justify the harshness and violence, the barbarous immediacy, that seem arbitrary in many of the others. He is a poet of great originality and power who has, extraordinarily, developed instead of repeating himself. His poems have a wonderful largeness and grandeur and exist on a scale that is unique today. You feel before reading any new poem of his the uneasy expectation of perhaps encountering a masterpiece.

The next morning the National Poetry Festival found itself in competition with the Cuban Missile Crisis. Attendance fell off dramatically since a number of poets felt that this could be The End and preferred to meet theirs at home. Rexroth luckily secured an earlier flight to San Francisco, the Nashes holed up in their hotel room with the television until Ogden's reading, and Mrs. Kennedy's sherry party was canceled—to my deep disappointment, for Mrs. Kennedy was my Princess Di. Instead of sherry at the White House I disconsolately went hunting for a bead curtain for Alleyne's room at college while Randall joined Mr. Ransom's discussion group. At the Customer Information Desk in Hecht's I stood by while the attendant dialed upstairs to inquire for the bead curtain, but when I heard a brusque male voice at the other end of the phone call answer, "The Pentagon," I beat a hasty retreat to the hotel. Frost spent the day in bed with shades drawn, Bill Meredith reported, but rumors flew this was not from fear of the bomb but from grief at losing the Nobel to Steinbeck's life work. A diminished group went to the Udall's party, Snodgrass among them, but Schwartz stayed in the hotel room drinking and smashing the furniture. As Delmore hurled the telephone out the window the police arrived and Berryman came to his aid. Together, and three sheets to the wind, they battled the foe through the lobby and into the paddy wagon: jailed for the rest of the festival.

A week later Randall, still in his poetry bubble, wrote Red:

> Actually Washington gave one no time to worry much over Cuba and this was a blessing. I like getting to talk about American Poetry. It was interesting hearing what the poets said and what they read, and how the audience responded; in general I just sat dully through the bad conventional poems, and was affected by the good ones. This may sound too good to be true, but it's true.

In 1963 Randall was awarded a sabbatical and we promptly ordered a Jaguar sports car XK–120 in desert gold, with orange leather seats "the color of a new football," Randall said. We were to take delivery at the factory in Coventry. He then mapped out our course across Europe on what a gentleman in Bamberg called our "secondt vedding moon." Randall had given England short shrift because he had fallings-out with Spender, Roy Campbell, and the Sitwells which had alienated him from the whole country.

In the lamp-lighted evenings at the Taylors' house Peter took to genteel frowning over our itinerary and venturing many a tactful "This won't do." He and Eleanor—such Jamesite, castle-combing, monarchistic Anglophiles—forced Somerset, Goathland, and Ross-on-Wye upon us with the same unreasonableness Randall had inflicted Freud, *War and Peace*, and Bosch on them. Later, I insisted he see Durham

Cathedral—the Romanesque beauty—for which he was eternally grateful and called me "Magic girl!" and himself "Dopey-old-Ramble-not-wanting-to-drive-out-of-our-way."

Hither and thither but *not* out of Randall's way were Hardy's house Max Gate, Wordsworth's Dove Cottage, Henry James's Lamb House, and Kipling's Bateman's. We made long, distant pilgrimages to each one, where, Randall not deigning to be a tourist, we sat reverently outside in the car before the shrine.

Randall couldn't get over his astonishment at the way London looked compared to New York. "Why, it's like a time machine," he said. Everything cozy and well made and stable that had fallen out of America in this century seemed to be going strong in England. He still liked Germany, but it was hard on him to be cut off from magazines and newspapers. "Oh, I'm so ill-educated," he fretted. "Imagine knowing *one* language!" The language we got fondest of in England we heard on a television program called *Steptoe and Son*, and Covent Garden had to carry on without us on their night. This and a curiosity about *cricket* made us feel somewhat residential. More and more, England seemed the fulfillment of Randall's wish for a foreign country where they spoke English. At last, one day over game pie at Fortnum and Mason's, he said, "We *ought* to get a flat here some summer." And it became a family saying like "California in the old days."

While the Wimbledon matches were televised, we set out early to do a gallery or a museum in the mornings. Then, propped in bed with our tea trays and sultana cake, we watched in bliss the mounting Australian sweep that year, before those hushed outdoor stands, with only the bup-bup cotton bubble sound that tennis makes.

In the long London twilights we liked to see the black and white parents in our district strolling in the park with their golden children. "English trees *are* like Constable trees," Randall noted interestedly when he looked out the window of our hotel on the Bayswater Road. His sightseeing continued to be a joke. No Westminster Abbey, no St. Paul's, no Tower, no Stratford-on-Avon; but we extended our stay. The next thing I knew, Randall had appointments on Savile Row with "this dovey tailor!" He even took me there and we browsed among the bolts of broadcloth, tweed, twill, pinstripe, and plaid, thinking of more raiment for him to order. One pale, almost rainbow-hued check kept drawing Randall back. Of course it was too *jeune fille* for a jacket, he said, and too overwhelming for a suit, and yet . . . and yet . . . When I suggested a topcoat, Randall burst out, "A topcoat! A summer overcoat! I'll be like someone in a Russian novel." He carried a stick for one day, but after he snapped off the tip in a grate he forgot it in a cab. When he remembered it, he said "Alas, poor stick," and laughed. "Well, it was hardly my style." A Burberry from Burberry's was. And a trilby from Locke's.

Randall wrote his mother, "We're crazy about England. The people in the parks and streets are awfully good humored and nice." He wrote Michael the opera and ballet were wonderful and wrote the Taylors, "We ought to get a flat here next summer. What are your plans?"

Cal was passing through London then and we spent a few hours together, talking on rented chairs along Rotten Row. Cal was for Plath that day, and Gunn and Larkin. Randall was for Larkin, Larkin, and Larkin. Cal's and Randall's temperaments were about as opposite as some of the poets they compared. At each visit the all-enduring friends roused each other up once or twice, whether they meant to or not; and then, like physicists on different hemispheres who advance their own knowledge on each other's papers, Cal and Randall, when their initial resistance passed, often pushed with their paws and found something palatable in each other's latest Enthusiasm. "Cal's right," Randall might say. "I was dumb about X. He's better than I thought." Or he might say, "The people Cal likes! Gee!" For a day or two after being with Cal, Randall was more Randall then ever.

His former phobia about England came down hard on France. Since he'd learned and *liked* to "maintain the orderly sequence of the queue," he was appalled at the brother-against-brother crowds in Paris. At *Tannhauser* one night where a line should have formed and didn't Randall said in a carrying voice, "Good *grief*! What a way to behave." One weekend did for France.

A few days among the hill farms of the Moselle and a few days of Alp viewing and miniature golf on the Boden See got us in the right mood for Munich. Mornings there started with a hectic routine: the Jarrells half through their breakfasts and urging on the half-asleep Michael di Capua with his, so we could be first at the box office for the turned-back opera tickets of the day. By ten-thirty the crisis would be passed. Either we were going that night, some of us were going, or none of us was going, and we could calm ourselves over *Weisbiers* on the sidewalk at Luitpold's. There Randall and I stared with tender fascination at a waitress who would never know she looked "like the perfect Gretchen for *Faust*." And there the now-awake Michael became, with his repository of opera knowledge, that rare kind of good company Randall so often longed for: "Someone who knows more than I do about a subject I'm interested in."

By ourselves in Vienna we had a month of choosing between the Theater an der Wien, the Staatsoper, and the Redoutensaal *every* night. Sometimes this could mean choosing

between Bergonzi or Seefried or Rysanek. For overall quantity
and quality the Kunsthistorische became our latest "unfash-
ionable enthusiasm." To have the nude Saskia and Titian's *Su-
sanna and the Elders*, Vermeer's incomparable *Artist in His Studio*,
and that Adam and Eve with the reptile that Randall called
"the one inspired Cranach," *and* Randall's beloved *St. Sebastian
Mourning St. Irene* was, to us, rich fare. The Velázquez Room
with, among others, nine Infantas made Randall cry out, "Oh,
Mama! The paintings Austria stole." And all this before you got
to the fourteen Brueghels. When we went into that room,
Randall whispered fiercely, "That Louvre! Isn't it *overrated*?" We
sat among the Brueghels often, and in silence got free of ourselves and into those scenes in water, in snow, in fields, where
the least crow flapping home was intended and where man's
small, coarse, mystical, frantic efforts in his world took place.
Before we left Vienna for a Donatello tour of Tuscany, Auden's
"Musèe des Beaux Arts" was rankling in Randall's brain and a
little later he wrote of his own feelings in "The Old and the
New Masters."

II

Home in Greensboro, Randall gave the finishing touches to his translation of *The Three Sisters* for the Actors Studio production. He was back to playing tennis on fine winter days. On other days he'd dress warmly in the ochre twill "Danish nobleman's jacket" and the briarproof trousers and lace-up Bean boots he is wearing in Betty Watson's portrait. He'd walk the Quaker farms near our house, finding winged elm branches and boughs of wild persimmon to bring home with the news of the ponies he saw and the little boy named David and the red dog that all figure in *Fly by Night*.

Randall liked us to have "cottage-y" meals, that is, omelet-and-salad meals with the teapot and the toaster on the table. But never anything white, like grits, mashed potatoes, or rice. He invented omelets with quartered radishes and avocado cubes in them and was deliciously enterprising with his salad dressings. Herbs and spices would be strewn about the counter in lavish disarray, but I was never sure which ones he'd used. Randall wasn't sure, either. Always flattered to be questioned he would helpfully try to remember and give up to say laughingly, "All that comes to me is what Frost said: 'A little of anything goes a long way in a work of art.'"

When the lettuce was especially nice, Randall sorted it over himself. Sometimes he'd call to me, "Peaches, come see your

Randall. He has something to show you. You'll be *glad* you came." And when I got there he'd show me a miniscule perfect lettuce leaf about as big as a canary feather, "Isn't it *dear*? I knew you'd want to see it." And he'd pop it in his mouth, saying, "It was much too good for this world."

Randall's youthful Marxist politics toward "this world" had pretty much withered away when he left *The Nation*. Hannah had displaced the Edmund Wilson and Philip Rahv clique. Europe, and especially the Rhine, had inspired his poetry and his novel to the point that for the next several years "Politics," as he himself said, "came off nowhere."

Kennedy changed all that. For Kennedy, we subscribed to the *Washington Post*. For Kennedy we moved up to a twenty-three-inch television screen where Kennedy even took precedence over Unitas. In a panel discussion for PBS Randall delighted in saying:

> I think all of us are grateful that for the first time in the history of the Republic, a great poet was invited to help with the inauguration of the president of the United States. Kennedy made his invitation not as a friend, not as a politician, but as a *reader*: any of us who heard the president talk about Frost's poetry recently will remember that he spoke as only a real reader of Frost could speak, and read the lines almost as Frost himself would have read them. It is good to have Fred Waring in the juke boxes, but it was sad to have Fred Waring, nothing but Fred Waring, in the White House, too. What a pleasure to think that for the next few years our art and our government won't be complete strangers.

Randall and I had summered in Santa Barbara when Luther Hodges, our North Carolina governor, nominated JFK at the convention in Los Angeles. Enraptured, we hung on Kennedy's every word, even touched by his shrill, almost adolescent treble in the acceptance speech that some adept voice coach soon managed to lower a half octave. Randall joked that JFK was the only man he knew whose voice changed in his forties.

How fondly attracted we were to catch on TV some of Kennedy's flights to Cape Cod (times stolen from the campaign), wearily deplaning, rumpled and hair blown, almost stumbling into Jackie's outstretched arms. And she so prominently pregnant.

When, as president, Kennedy attended the University Founder's Day Celebration in Kenan Stadium, Randall and I drove to Chapel Hill and watched him through binoculars from the fifty-yard line marching in the October sunlight talking amicably to President Friday—of the university—and those around him and making everyone laugh.

Throughout the weekend of the assassination, we two shared boxes of tissues in front of the television insatiable for the events of Kennedy's boyhood; Irish-rough family football; the war heroism; the back surgery and Jackie's devotion; and his Democratic career from senator to president.

However, in the wake of the tragedy when the *New York Times* telephoned asking Randall to write a poem for Kennedy, it was like the bat-poet trying to write about the cardinal. Impossible. Pillowed in our bed with our ailing cat, Elfie, curled beside him, Randall got as far as "The shining brown head . . ." and again "The shining brown head . . ." before crumpling the page time after time and pitching it accurately into the wastebasket.

The Taylors were not as stricken as we were. In fact it seemed best not to speak of our feelings with them, or with Cal, about Jackie's solitary visits to the Rotunda late at night to kneel by and rest her head on the casket. Or Caroline's telling her nursery school teacher, "My mommy cries all day."

A month later Randall was still grieving and wrote Michael, "President Kennedy's death to us was the saddest public thing that's happened in our lifetime, and we can't really think about anything else."

The spring of 1964 was melancholy for us, especially for Randall. Our flawless new XK–120 developed knocks and clunks no local mechanic knew what to do about. Our dearly loved Elfie slowly succumbed to leukemia, and the Pulitzer Prize went to Phyllis McGinley—thanks to Robert Frost. "What a world," said Randall, aggrieved, "People! *People!*"

Then there was Michael's sudden departure from Macmillan, which was extremely unsettling for Randall since he was relegated to a strange new editor for his next three books and the forthcoming *Bat-Poet* and had lost the empathy and judgment of this trusted friend who had helped him break his writer's block and who was the whole reason for his going to Macmillan. As usual Randall tried to cheer himself—and to cheer Michael while he sought another publishing house. But he was shaken. On the plus side, Randall accepted an invitation from Kenneth Tynan to read poetry at the Edinburgh Festival in the summer of 1965. However, his fiftieth birthday made him feel old, and added to this, his mermaid was a grandmother. We did anticipate the opening night of *The Three Sisters* and the Shubert's Alley reception, and Randall wrote Cal: "Geraldine Page is Olga, Kim Stanley is Masha and Shirley Knight Irina. If only they get a nice but ugly Tuzenbach." We'd expected to stay in New York and see the play several times but Randall was totally disappointed. "Except for Masha," he wrote Cal. "But she was much too fat, poor thing. It was a disaster. As crude and exaggerated as Chekhov always is in this country." Knowing exactly how *he* would have acted them, Randall was merciless toward Vershinin, Dr. Chebutykin, and Tuzenbach. We left New York the next morning.

The inertia-fatigue that dogs the depression victim was invading Randall's reading and writing. His mother told him he was making mountains out of molehills. He told his mother, "When you're depressed there are no molehills." Then came his fall in a doubles match, when Randall dove for a low, running forehand, shouting, "Mine! *Mine!*" and crashed on the concrete court, sharply twisting his sacroiliac and sidelining himself from tennis for weeks.

His melancholy deepening, he scorned the words "midlife crisis" and kept on a slip of paper in his wallet the Freudian term "*torschlusspanik,*" that is, door-closing panic. Soon we sought help in another state from a psychiatrist who had treated Cal.

This doctor knew at once that Randall was not a schizophrenic like Cal. He promptly prescribed a newly marketed mood-elevating drug, and we drove back to our Blue Ridge mountains for the rest of that summer.

Shortly after this a Coventry-trained mechanic restored the Jaguar to "mint condition." Michael and *The Animal Family* signed on with Pantheon, and Randall wrote his first poem in over a year, "The Player Piano."

At the time "The Player Piano" was finished, my mother was visiting us and she and I and the cat, Elfi, were Randall's first audience. My eighty-year-old mother sparkled with joy throughout his reading and declared in astonishment, "Oh, I *understand* this one, dear boy. I *really* do . . . The false armistice! Mary was just a little tyke and we were living with my mother in Jefferson City while her father was overseas. I remember it all perfectly . . . oh, and *Fatty* Arbuckle. We were in Pasadena when Fatty Arbuckle drove the El Molino bus." Then she said the sweetest words a poet ever hears, "Won't you read it again, dear?"

The Player Piano

I ate pancakes one night in a Pancake House
Run by a lady my age. She was gay.
When I told her that I came from Pasadena
She laughed and said, "I lived in Pasadena
When Fatty Arbuckle drove the El Molino bus."

I felt that I had met someone from home.
No, not Pasadena, Fatty Arbuckle.
Who's that? Oh, something that we had in common
Like—like—the false armistice. Piano rolls.
She told me her house was the first Pancake House

East of the Mississippi, and I showed her
A picture of my grandson. Going home—
Home to the hotel—I began to hum,
"Smile a while, I bid you sad adieu,
When the clouds roll back I'll come to you."

Let's brush our hair before we go to bed,
I say to the old friend who lives in my mirror,
I remember how I'd brush my mother's hair
Before she bobbed it. How long has it been
Since I hit my funnybone? had a scab on my knee?

Here are Mother and Father in a photograph,
Father's holding me. . . . They both look so young.
I'm so much older than they are. Look at them,
Two babies with their baby. I don't blame you,
You weren't old enough to know any better;

If I could I'd go back, sit down by you both,
And sign our true armistice: you weren't to blame.
I shut my eyes and there's our living room.
The piano's playing something by Chopin,
And Mother and Father and their little girl

Listen. Look, the keys go down by themselves!
I go over, hold my hands out, play I play—
If only, somehow, I had learned to live!
The three of us sit watching, as my waltz
Plays itself out of a half-inch from my fingers.

The Complete Poems

When the long-distance psychiatrist telephoned to check on Randall's state, Randall had nothing but good to say of his "magic pills" and how well they were doing their job. Had the doctor actually seen Randall, I feel sure he would have cut back on the medication and let Randall stabilize. However, he renewed the prescription.

Before long the depression was forgotten. Everything Randall's heart desired seemed possible. He played tennis until dark and played his high-fidelity until dawn. Almost. The State Department scheduled him for a Cultural Exchange trip to Russia. He met Unitas on an airplane and wrote a poem about it the next day. He was so energized that he barely slept and barely stopped talking and poems flew at him like never before: quatrains, couplets, haiku, parts of poems, and ideas for poems. Still on the "magic pills" Randall quoted Shakespeare's *Richard III* to us, so gaily, "Richard is himself again."

Though the students and younger faculty were enthralled, Randall was no self his nearest and dearest had ever known. Fi-

nally, at the instigation of the English Department, Jim Ferguson, the new chancellor, benevolently escorted Randall to the
University Hospital in Chapel Hill, "for a rest." Before it was
finished with us the ordeal called forth a desperate valor we
would never have known we had.

By July 1965 Randall was home again and our lives were
turning right side up. While the Edinburgh Festival had to be
canceled the Russian trip was simply postponed, and a writer-
in-residence year at Smith College was contracted for 1966–67.

In our Indian summer, after Randall's classes, we often took
our binoculars and our Bicycle cards and drove out in the great
world to Piney Lake. Between swims and bird watching, we
played an innovative pinochle Randall invented and we kept a
running score. Other afternoons we lay outside in our twin
hammocks while Richter played Chopin Preludes with the
volume turned high and we read our twin copies of *The Inspec-
tor General*, which Randall was teaching. Time and again he
looked up and laughed aloud. "Why in the world doesn't our
Drama Department put this on? I could play Danny Kaye's part
for them in my sleep."

The cloudless blue skies had one of us repeating to the
other—as we always did in September and October, "It's like
California in the old days." We drove to a farm on Horse Pen
Creek Road, where we picked scuppernong grapes and accepted the farmers' annual invitation to "set a spell" and cool
off with a glass of water from "the best well in the county." In
the mornings I would wash and boil the grapes and set them
dripping from a jelly bag hung above a homely crock I kept for
that purpose. Randall relished this little harvest ritual, which
made him feel in touch with "real country" people. At some
point when he joined me in the kitchen, he would greet my
Rube Goldberg device, "Well, hail to thee, old jelly bag! And

welcome back!" He especially liked for me to display our winter supply of paraffined jellies on the counter for weeks for visitors to see and to be replaced by my dark German fruitcakes ripening in Courvoisier inside their foil wrappers until Christmas.

I was usually the one to go inside and change the record and check on the spice cake with grated lemon rind and baked-on seafoam icing that was Randall's favorite. Often I came back to him concentrating on the music with his eyes closed and playing chords on his solar plexus. We listened to the Preludes and nothing else for weeks, marveling at the mood-pictures of tenderness, ecstasy, melancholy, patriotism, and grief that Chopin built out of Bach and that Liszt would build romanticism out of Chopin.

We had exhausted our own analyses of Randall's nightmare experiences some time before and the last reference he made to me came while we sipped new cider by a blazing maple in our woods "having its day" as we said. I was holding a red and yellow spray of leaves to take inside when Randall said, apropos of nothing, "It was so queer, beloved. As if the fairies had stolen me away and left a log in my place."

In a letter to Maurice Sendak, sympathizing with his misery over his elderly father's critical illness, Randall wrote:

> I hope the days with your father will be better now. Our misery about such things is so hard to explain or bear—I guess all one can say is that we can't escape them, and that afterwards we know what matters most to us, what life really seems to us, and are better off in that way, we know that the surface that seemed to matter to us is just the surface, and that it doesn't matter compared to our real life and real self.

April 1965

On April 18, 1965, Randall, deeply despondent, read on a back page of the *New York Times Book Review* section a short review of *The Lost World*, written by Joseph Bennett, editor of the *Hudson Review*.

Lumped with more favorable reviews of Samuel French Morse, Hollis Summers, and Hayden Carruth, Bennett wrote:

> Randall Jarrell's "The Lost World" contains four poems of considerable interest where the craftsmanship is clear, the intention honest and skillfully realized in the work: "Bats" with their "needlepoints of sound;" "Woman" serenely and confidently written; "In Nature There Is Neither Right nor Wrong" and the uncannily atmospheric "A Hunt in the Black Forest," the only first-rate poem in the volume.

> With the exception of the poems mentioned above, the rest of the book is taken up with Jarrell's familiar, clanging vulgarity, corny cliches, cutenesses, and the intolerable self-indulgence of his tear-jerking, bourgeois sentimentality. Folksy, pathetic, affected—there is no depth to which he will not sink, if shown the hole. Cultural name-dropping, hand-cranked puns and gags—a farrago of confused nonsense, a worn-out imagination. There is even a stage Italian out of Hollywood, placed incongruously in the Museo delle Terme; it is all Very Cultured. Jarrell's stance is the fashionable anti-intellectual one of fifteen, twenty years ago. His work is trashy and thoroughly dated; prodigiousness encouraged by an indulgent and sentimental Mama-ism; its overriding feature is doddering infantilism.

Within days of this, and in such unrelieved depression that shock treatments were being considered, Randall cut his left wrist in a suicide attempt. Ironically, this act seemed to release him from his depression. When he sent word to me, I wrote back, "Dearest One, I think of you constantly and miss you so much. How I wish we could be back with each other. I hope your poor wound doesn't hurt so much now. I don't know what else to say except I long to have you well and your wonderful self again. I'll telephone as soon as they permit it. God bless you and keep you for me. You're all I want."

[May 1965]

Dearest:

It's Sunday morning, and I've just looked at the roses you brought me from our own front yard, and put on the nice shoes you brought me—and now I'm writing this on the clipboard you brought me, and thinking (just as I did lying in bed) of all the other things you brought me that for so long made me better and happier and kept something like this from happening to me earlier. You are the one big good real thing in my life, and I'm so glad I met you—we've had hard times together, but so much more happy and good time, time different from any other in our lives.

I've been remembering our first days in Boulder, and Denver, and Grand Lake, and how you looked in that yellowish sweater of mine, and all the time at Princeton—writing so much and talking to you long distance, and saying good-bye at the airport. And how wonderful it was visiting you at Christmas. And Yosemite after we were married, and the way it looked driving to Sequoia on the way back.

I love you,
Randall

Beatrice was starting her junior year at North Carolina State University.

[September 1965]

Dear Beatrice:

This is a start-of-school letter from a beginning professor to a beginning student. I had my first two classes yesterday—one about Turgenev and one about Hopkins—and it all felt so familiar and nice; as Mary and I drove home it was as if I'd never been away.

We had a nice party the night before for the Joseph Bryants, [Head of the English Department], the Watsons, and the Head of the Art Department [Gilbert Carpenter]. The William Morris Scarf was framed and hung on the wall, the new living room was all in its glory, and it was the prettiest party we've ever had. I'm certainly glad you suggested putting the Renoir lithograph over the fireplace; it looks so beautiful and becoming there.

What with seeing you when you came over I haven't been writing you—but I did want to send you those little to-be-used-for-frivolous-things checks that I mentioned when I was talking to you a couple of months ago.

I hope your new classes are as nice as mine and that everything's as nice in Raleigh as it is in Greensboro. See you soon, darling.

Love,
Randall

(Note attached to a wooden birdhouse Jarrell gave to me)

October 1965

To my Mary:

Each of us is the other's bluebird of happiness, but it's nice for us to have, for any wandering bluebird in the world, a little brown house in a tree. Remember the day when the whole flock of bluebirds spent the afternoon in our front yard? I hope they do again this fall . . . I've just been re-reading some of the letters you've sent me and I felt very happy and lucky to have someone love me so and write me so. I love you.

Your Randall

Although I regularly massaged Randall's wrist and he had not missed a day flexing it, soaking it in hot water, and squeezing a rubber ball in his fist, it was not responding as it should but seemed to be stiffening and pulling his hand down. A hand specialist in orthopedics at the University of North Carolina medical school recognized the problem at once, saying, "Oh yes, wrist drop or Pollock's wrist," and he explained how a Dr. Lewis Pollock had studied thousands of such peripheral nerve injuries and treated them successfully with intensive physical therapy and stimulation. The orthopedist prescribed this for Randall as part of a daily regimen at the Hand House of the university medical school when space was available and assured him the prognosis was good.

On Thursday, October 14, about 7:30 in the evening, Randall wearing his dark wools and dark gloves, was walking along the edge of the highway on his way back to campus. He was not visible to the car's driver until too late to be avoided, and in a split second he was sideswiped—"hit from the side," as the coroner said, "not the front, or the front wheels of the car."

The Chapel Hill newspaper reported that "the impact spun Randall around and knocked him not more than three or four feet on to the grass by the edge of the road."

Questioned, when the patrolman arrived, the occupants of the car first said, "He seemed to whirl," adding afterward that "he appeared to lunge in the path of the car." Had Randall "lunged in the path of the car," he would have been run over, but that was not the case. The coroner found he was hit by the side of the car, not the front or the front wheels. This agrees with the reporter's statement that the impact spun Randall around and knocked him to the side, as would have happened since he was vertical. Further evidence of this is shown by the injuries described in the autopsy. All Randall's injuries were sustained on one side, from the abrasions on his cheek and the fatal blow at the base of his skull to the fractured foot, which was the only part of his body that came in contact with a wheel. Being vertical he hit and damaged the side rearview mirror.

Although the newspaper reported that "a two-day old, un-filled pain-killer prescription" was found in Randall's wallet to be used in case of extreme aching from the physical therapy on his wrist, the prescription had not been filled and the autopsy disclosed "no evidence of intoxication or any other disease process which might have contributed to his demise." These medical findings, plus a thorough three-week investigation of the circumstances, brought the doctors in charge of the autopsy to the conclusion stated for the media, that there was "reasonable doubt about its being a suicide." In the judgment of the coroner and the medical examiners—all of whom had seen the body—Randall's death was accidental, and that is what is given on the death certificate. Not a suicide as was precipitously rumored in some quarters.

In my memories Randall is always stylishly outfitted. And he was that fateful night in the October cold of the early dark in Chapel Hill, 1965, wearing his dark Savile Row Shetland jacket, a dark tweed cap, and his dark Italian leather driving gloves. The innocent motorist declared he "never saw him until it was too late."

My last memory of being in Randall's company was a dazzling fall Sunday, one of those days when the moon can be seen in the sky along with the sun. We'd spent a loving afternoon by ourselves and he could no longer put off packing the little straw picnic hamper he currently preferred to "real" luggage. I sat in the rocker in our bedroom as Randall playfully tossed in a shirt or two and a sweater, telling me, "I can buy anything else over there at that dovey men's shop."

Then he piled on the three-volume variorum edition of Emily Dickinson's poetry with his notes for a critical essay and after that his review copy of Elizabeth Bishop's *Questions of Travel*.

Outside on the porch steps we posted ourselves waiting for Bob Watson's arrival to take Randall to Chapel Hill for a period of physical therapy on the wrist he had despairingly cut earlier that spring after a scathing review of *The Lost World* in the *New York Times*.

Mainly, Bob simply wanted to be with his good friend, but partly, they saw it as an opportunity to go over Hopkins's somewhat esoteric "Wreck of the Deutschland," which Bob was to teach to Randall's class while he was gone.

When Bob drove into the driveway Randall and I kissed unabashedly and Bob smiled and tried to look off somewhere. Then Randall sped over the flagstones, flung the hamper into the car, and was about to climb into the front when the idea struck him to run back where I was standing under a big oak and throw his arms around me for one more hug and kiss.

Randall is buried near our home in the pre-Revolutionary Quaker cemetery of the New Garden Friends. A large pin oak spreads its branches over him and mockingbirds sing from the treetop. The words engraved on the upper half of the flat, Moravian-style ledger are:

RANDALL JARRELL

Poet

Teacher

Beloved Husband

1914   **1965**

The lower half of the eight-foot, horizontal rectangle is left blank, biding the day when my name is carved in the stone with his for the rest of time.

Today, engraved in my memory for these thirty-four years, is the solace of the service for the Burial of the Dead at Holy Trinity Episcopal Church. At will I can retrieve the anguished faces of the pallbearers: Peter, Cal, Bob, Michael, Otis Singletary, and Bill Carrigan, Randall's tennis partner.

The strains of Bach's "And Sheep May Safely Graze" always comfortingly remind me of the penitential purple mantle over the casket that I sat beside with Randall's family of women: with Irene, my mother; and Alleyne and Beatrice, my daughters, all of us in black. And I still say to other bereaved the words a former Eliot Girl said to me that day: "Death ends a life but not a relationship."

AFTERWORD

In the days that followed, Cal's book *The Old Glory* arrived, dedicated to Randall and to Jonathan Miller and inscribed: "I am heartbroken Randall cannot have this book he loved and helped." A note came from Peter, grieving over the recent loss of his brother-in-law and his father as well as Randall; he wrote, "My brother-in-law introduced me to Chekhov; Randall taught me how to read Chekhov; and my father showed me what Chekhov meant. I loved all three." Bob Watson said—and continues to say, "I can't realize he's gone. I can still hear his voice, his words clear and shining."

Author's Note

Inspired by Sir Kenneth Clark's disclaimers for his memoirs written across several decades, as mine are, I, too, as he said,

> "Have recorded the events and moments that have remained most vividly in my mind or that still make me laugh when I recall them." I, too, never kept a diary. "In consequence," Sir Kenneth went on, "the narrative may contain mistakes of chronology and other inaccuracies, but one thing I can guarantee—the complete accuracy with which I have recorded the spoken word. Some of the sentences I quote may seem improbable but I could still reproduce the tone of voice in which they were said."

And recall the rhetoric, the smile or scowl; the tenderness or outrage that went with them.